OXFORD MEDICAL PUBLICATIONS

Aneuploidy

OXFORD MONOGRAPHS ON MEDICAL GENETICS

General Editors:

J.A. FRASER ROBERTS
C.O. CARTER
A.G. MOTULSKY

OXFORD MONOGRAPHS ON MEDICAL GENETICS NO. 11

Aneuploidy

D.J. BOND
Department of Genetics,
University of Edinburgh

and

ANN C. CHANDLEY
MRC Clinical and Population Cytogenetics Unit,
Western General Hospital,
Edinburgh

OXFORD   NEW YORK   TORONTO
OXFORD UNIVERSITY PRESS
1983

Oxford University Press, Walton Street, Oxford OX2 6DP

London Glasgow New York Toronto
Delhi Bombay Calcutta Madras Karachi
Kuala Lumpur Singapore Hong Kong Tokyo
Nairobi Dar es Salaam Cape Town
Melbourne Auckland

and associates in
Beirut Berlin Ibadan Mexico City Nicosia

Oxford is a trade mark of Oxford University Press

British Library Cataloguing in Publication Data

Bond, D.J.
Aneuploidy.—(Oxford monographs on medical genetics)
1. Aneuploidy
I. Title   II. Chandley, Ann C.
574.2   QH461
ISBN 0–19–261376–6

Library of Congress Cataloguing in Publication Data

Bond, D.J.
Aneuploidy.
(Oxford monographs on medical genetics) (Oxford medical publications)
Bibliography: p.
Includes index.
1. Aneuploidy—Addresses, essays, lectures.
2. Human chromosome abnormalities—Addresses, essays, lectures. I. Chandley, Ann C.
II. Title III. Series: IV. Series: Oxford medical publications. [DNLM: 1. Aneuploidy.
QH 461 B711a]
RB155.B615   1983   616'.042   83–2195
ISBN 0–19–261376–6

Set by Promenade Graphics, Cheltenham
Printed in Hong Kong

Preface

In most genetics textbooks the subject of aneuploidy receives relatively little attention. In some it is afforded a section of a chapter; in others, a whole chapter might be found. Rarely, however, is the subject considered in any depth. And yet, as a topic which impinges on man it is of considerable direct interest because the frequency of aneuploid conceptions in our species is very high and the resulting burden, both financial and emotional, is not inconsiderable. In man the underlying cause of the dramatic increase in aneuploid conceptions among older women remains the most important unanswered question in this area of research.

When the suggestion was first made that we write a book on aneuploidy it seemed timely, not only because of the direct relevance of the subject to man, but because of the growing realization that exposure to environmental mutagenic hazards might increase this already high frequency. The literature was, however, scattered widely in a diverse array of publications, dating back to the year 1921 when James Mavor first reported his findings bearing on loss of the X-chromosome from the oocyte of *Drosophila melanogaster* using X-irradiation.

In writing the book, our aims therefore have been to try, first and foremost, to draw together some of the multifarious threads. We give consideration to the problem of human aneuploidy, its frequency at conception and at birth, the mechanisms which may operate to generate it both spontaneously and following environmental insult, and its consequences for the individual or family. We survey the data from a range of species from fungi to non-human primates, taking into consideration the spontaneous incidence of aneuploidy and the tests which have been carried out to induce aneuploidy by experimental means. We try to reach some conclusions about the relevance to man of testing for aneuploidy in lower organisms and especially in non-mammalian species. Finally, we consider the future and try to point to the areas requiring further investigation and a more detailed consideration.

We hope that our readership will include not only experimental researchers interested in the basic mechanisms underlying aneuploidy production and environmental mutagenesis specialists interested in aneuploidy testing, but also human cytogeneticists and clinical geneticists interested in the aneuploidy problem as it relates to man. Our aim will be to put the human aneuploidy problem into perspective and provide a framework for the future development of a complex area of research.

Edinburgh   D.J.B.
January 1983   A.C.C.

Acknowledgements

Several colleagues and friends, both in the University of Edinburgh, in the MRC Clinical and Population Cytogenetics Unit, Western General Hospital, Edinburgh, and elsewhere, have read parts of the manuscript and have greatly improved it by their helpful suggestions and criticisms. In particular, we would like to acknowledge the contributions made in this respect by Dr A.J. Bateman, Professor Max Clark, and Professor C. Auerbach. Special thanks are due also to Mr N. Davidson for his skilful execution of the text figures, and to Mrs S. Lochrie and Mrs A. Kenmure for the typing of the final manuscript. The contribution of Mrs N. Webb and Mrs J. Bogie to the typing of the early drafts is also acknowledged. Miss S. Mould, Librarian, is thanked for her invaluable assistance in obtaining necessary references and reprints.

Finally, we are indebted to Dr E.P. Evans and Miss A.M. Fulton, for supplying us with unpublished data and to Mr J.D. Brook for allowing us to use his unpublished illustrations (Figs. 5.1a, b, and c; Fig. 5.2; and Figs. 5.3a, b, and c).

Contents

Plates fall between pages 54 and 55

1. Introduction

There is probably no known single genetic hazard of more importance to man than the creation of imbalance by nondisjunction or chromosome loss leading to aneuploidy. More than 70 per cent of the chromosomal errors associated with human early spontaneous abortion and nearly one half of all those detected among newborns are aneuploids.

Aneuploidy is a condition in which an organism or a cell possesses fewer or more chromosomes than an exact multiple of the haploid number. It is a term which was originally introduced by Täckholm (1922) following his studies on the Rose genus. When an organism has one more than the diploid number of chromosomes, it is termed a 'trisomic', when it has one less than the diploid number, it is termed a 'monosomic'.

In mammals, aneuploidy usually has adverse effects on the individual concerned, and in man, the consequences can be severe in terms of human suffering and unhappiness. There is no known liveborn monosomic or trisomic condition which does not confer some phenotypic disturbance on the individual carrying it. For many of the liveborn sex-chromosome aneuploids, there is an uncertain prognosis and an increased risk of emotional, physical, and intellectual problems in later life and for most liveborn autosomal aneuploids, there will be mental incapacitation.

This detrimental effect on the organism comes about through changes in gene dosage and genetic imbalance which arise as a result of the change in chromosome number. The important feature is not the abnormal total number, but the identity of the missing or extra chromosome. Since each chromosome contains a unique collection of genes, the precise phenotypic consequences of the genetic imbalance will depend on which genes are extra or deficient.

An aneuploid individual usually arises at fertilization by the fusion of an abnormal gamete, which in turn has arisen through a defect in the meiotic division which gives rise to that gamete. The error can arise in either oogenesis or spermatogenesis and at either the first or second meiotic division. The complexities of the process will be presented in detail in the body of the book, but the most frequent final event is the failure of a homologous pair of chromosomes to segregate so that instead of one going into each gamete, two go into one leaving none for the other. The most usual process giving rise to aneuploidy is 'nondisjunction'. Thus, monosomy $(2n - 1)$ and trisomy $(2n + 1)$ (the products of fertilization of nullisomic and disomic gametes respectively) are complementary consequences of nondisjunction. Monosomy, however, can also arise independently of trisomy, independently of nondisjunction, by an entirely different event, i.e. chromosomal loss. Thus, whereas trisomy is the hallmark of prior nondisjunction, monosomy is not. Other less common mechanisms may also operate to promote aneuploidy (see Chapter 3).

Man appears to have an unusually high frequency of aneuploidy (although this characteristic might be a feature that is common to primates in general—see Chapter 2) and it is possible that man-made hazards, to which human populations are increasingly exposed, will generate even more aneuploidy (see Chapters 6 and 7). In 1921, Mavor discovered that X-irradiation could induce aneuploidy in *Drosophila melanogaster*. This was a finding that preceded by several years Müller's discovery of the mutagenicity of X-rays (Müller 1927), for which he was subsequently awarded a Nobel Prize. Since these pioneering experiments, the informed world has become much more aware of environmental genetic hazards. The recognized sources of such hazards have been steadily extended, from the testing of nuclear weapons, to nuclear energy, medical uses of X-rays both for diagnostic and therapeutic purposes and the many radioisotopes which are byproducts of the nuclear energy programme. In addition, many classes of chemicals, originally referred to as 'radiomimetics' because they mimicked the biological effects of ionizing radiation, have been identified. Mutagenicity is a property common to many groups of chemically-active substances used both in the chemical industry (alkylating agents), in medicine (antibiotics and chemotherapeutic agents), and in agriculture (herbicides and pesticides). Indeed, the multiplicity of mutagenic chemicals, and the complexity of the chemical industry have given rise to a new field of activity—mutagenicity screening, which has the objective of testing old and new chemicals for the possibility of their mutagenicity, and assessing their potential as a genetic hazard. Not only is there the choice of which of the forms of mutation (and aneuploidy is only one) to test for, but on which organism to carry out the test. Theoretically, any species could be used (even man, whose protection against these hazards is our primary concern). Every geneticist, according to his experience, has his own favourite method. In general, the testing methods for aneuploidy are the most laborious and inefficient. Nevertheless, a good deal of experimental testing for aneuploid induction using ionizing radiation, chemicals and other compounds has been carried out, not only in species like *Drosophila*, but also in an assortment of mammals. The suggestion has even been made that a compound which is identified as an aneuploidy-inducer should be named a 'trisomigen' (Bateman, A.J., International Congress of Environmental Mutagenesis, Edinburgh 1977). This term was proposed in recognition of the fact that, as far as man is concerned, trisomy imposes the greater load on society. It also accommodates the possibility that some substances might be discovered which are not mutagens but which induce aneuploidy as their sole genetic endpoint. We will use it throughout this book.

What seems certain is that although experimental work in a number of species has thrown some light on the possible mechanisms involved in the production of aneuploidy, and on the exogenous factors which may enhance its production, little is known or understood of the cause of the high spontaneous level of aneuploidy in man, and the striking maternal age-related increases which occur for many of the small chromosome trisomics. Indeed, more has probably been said and written in the last twenty years about the aetiology of human aneuploidy, and yet less is probably known about this particular aspect of the problem, than any other. Whether the explanation can be found in some major biological difference between man and other species or whether aneuploidy in man is just another 'disease of civilization' awaits clarification. Current data from Denmark and several other countries are indicating an increase in cases of Down syndrome which are *not* related to maternal age and the suggestion is being made that environmental factors, particularly in urban areas, might be operating to promote extra nondisjunction in males as well as in females. Whatever the explanation, the fact is, that in spite of a declining mean maternal age for Down syndrome births over the past twenty years, the incidence of the condition appears to be remaining constant.

If this monograph helps to focus attention on the problem and points to areas in which more investigation is required, it may make some contribution to the reduction of a major source of human unhappiness.

2. Aneuploidy in man: the problem stated

In this chapter, an attempt will be made to assess the problem of human aneuploidy by summarizing what is currently known about its frequency. By stating the incidence of aneuploidy in human gametes and the contributions made by monosomics and trisomics to the total number of conceptions in man, the resulting genetic load created by this one class of chromosomal error alone will be highlighted. It will be seen that the estimated frequency of aneuploidy at conception is far greater than the frequency at birth, the intervening gestational period accounting for the selective death, by spontaneous abortion and stillbirth, of the vast majority of aneuploid conceptuses.

THE FREQUENCY OF ANEUPLOIDY IN HUMAN GAMETES

Direct estimates of the levels of aneuploidy in human gametes are based solely on studies which have been made in males. The technical problems involved in obtaining human oocytes in the required post-segregational stages for chromosomal analysis of the products of meiotic disjunction has precluded such studies in females to date.

In males, two approaches have been tried. The first is based on the analysis of double 'bodies' or 'spots', representing differentially stained regions of specific chromosomes, in the nuclei of spermatids or ejaculated spermatozoa. Estimates of the spontaneous meiotic nondisjunction frequencies of chromosomes 1 (Geraedts and Pearson 1973), 9 (Bobrow *et al.* 1972; Pearson 1972; Pawlowitski and Pearson 1972) and the Y chromosome (Pearson and Bobrow 1970; Sumner *et al.* 1971; Pawlowitski and Pearson 1972; Kapp 1979) have been made in this way. Using incubation in CsCl as a pretreatment, Meisner *et al.* (1973) were able to stain prominently the secondary constriction of human chromosome pair No. 1 in lymphocyte metaphase preparations, and when spermatozoa are treated by the same technique, a single large stained body is seen if segregation of chromosome 1 at meiosis is normal; two such bodies indicate nondisjunction for chromosome 1 (Geraedts and Pearson 1973). A frequency of between 1 and 2 per cent of all spermatozoa from one individual showed two such bodies. For chromosome pair No. 9, using a 'Giemsa 11' staining technique in which the secondary constriction is represented by a red-staining spot in the sperm head (Bobrow *et al.* 1972), and for the Y chromosome, using fluorescent staining by which the distal portion of the long arm is seen as a bright spot (the 'Y'- or 'F'-body) in the sperm head (Pearson and Bobrow 1970) estimates of 2 and 1.26–1.40 per cent respectively have been obtained by various authors (Pearson and Bobrow 1970; Sumner *et al.* 1971; Bobrow *et al.* 1972; Pearson 1972; Pawlowitksi and Pearson 1972; Kapp 1979). Thus the estimated frequencies of aneuploidy for chromosomes 1, 9, and the Y are similar when this method is used.

It has been argued (Pawlowitski and Pearson 1972) that if the behaviour of these three chromosomes were representative of all chromosomes in the human genome, an overall incidence of about 40 per cent aneuploidy for human sperm would be obtained. This would seem intuitively to be an extremely high and unrealistic value; it is based on the assumption that all chromosome pairs undergo nondisjunction at the same rate. This may indeed be the case, but on the other hand, it is possible that the factor which allows the special staining of chromosomes 1, 9, and the Y chromosome (i.e. the possession of large blocks of heterochromatin which may have a specific chemical composition (Jones and Corneo 1971)), may also predispose them to undergo nondisjunction at a higher rate than any other chromosomes in the genome (Geraedts and Pearson 1973). Ford and Lester (1982) have recently demonstrated, for example, that chromosomes with large C-bands are more often displaced from the metaphase plate of somatic cells than are those with small C-band regions. This may be an important factor in determining the predisposition of a specific chromosome pair to undergo nondisjunction, both at mitosis and meiosis. The precision of the data obtained from stained whole sperm has, however, been questioned on account of biasing factors which may enter into the scoring procedures (Roberts and Goodall 1976; Beatty 1977). Beatty (1977), for example, has suggested that some Y chromosomes may not be represented by F-bodies in sperm heads, while other fluorescent bodies, the so-called 'adventitious' bodies, may be mistaken for true F-bodies. Beatty's arguments are supported by DNA estimates of Sumner and Robinson (1976) which indicate the unlikelihood of a one-to-one correspondence between 2F and YY, many of the 2F sperm being 'not YY-bearing'.

It would seem therefore at present that F-bodies do not provide a useful basis for estimating the incidence of Y chromosome aneuploidy since the problem is overshadowed by complicating factors which are not fully understood. Furthermore, it is likely that the same criticisms would apply to the methods for estimating the nondisjunction rates of chromosomes 1 and 9. Because of the dubious results obtained from 'double spot' analysis of human spermatozoa, other methods have been sought. The most recent and exciting development is a method using the *in vitro* fertilization of Syrian hamster eggs by human spermatozoa with activation of the human sperm to a point where their chromosomes can be viewed directly (Rudak *et al.* 1978; Martin *et al.* 1982). The environment provided by the cytoplasm of the mature oocyte seems to be necessary for the reactivation of the sperm nucleus and subsequent decondensation of the chromatin (Yanagi-machi *et al*. 1976). By the use of suitably controlled culture conditions, discrete haploid sets of both hamster and human metaphase chromosomes, with up to three sperm complements per egg, can be obtained (Rudak *et al*. 1978). The sperm chromosomes appear large, with parallel chromatids in which distinctive coils can be seen (Plate 2.1) and by the application of banding techniques, the analysis of human sperm genomes can be carried out with at least the same precision as is possible in the analysis of chromosomes from human somatic metaphases.

Of 60 sperm analysed by Rudak *et al*. (1978) from one male, 31 showed a 23,X constitution, 26 a 23,Y constitution, and three an aneuploid complement. One of the latter cells lacked a G-group chromosome, one lacked an F-group chromosome, and one had gained an acentric marker. The frequency of aneuploidy for this small sample of human sperm was therefore estimated to be 5 per cent. Further, more extensive studies, using this technique, have since been carried out by Martin *et al*. (1982) and the most recent results (Martin, personal communication) provide data for the chromosome complements of 948 spermatozoa from 31 normal men. The overall frequency of abnormalities in this population of spermatozoa was 82/948 (8.6 per cent) with 5.2 per cent (49/948) aneuploid and 3.4 per cent (32/948) showing a structural abnormality or chromosome break. All chromosome groups were included among the aneuploid complements, but there was only one complement (22, -C or Y) that could give rise to a 45,X embryo. Of interest was one 24,Y + 1 (no trisomy 1 has yet been reported among spontaneous abortions (see below) and no 24,YY complements. Numbers of sperm complements were too small to assess individual variation, but it seemed that the normal range was 0–14 per cent abnormalities with two men showing much higher frequencies of 28 and 31 per cent. These two men had a high frequency of breaks.

The encouraging results obtained in this study show that the method is feasible for further surveys of individuals within the general population and within selected groups, and it is hoped that more information will be gathered by its wider application in the future. Indeed, the data obtained so far give encouragement to the view expressed by Ford (1975) that, given at least an equal contribution from the male and female partner, a chromosome abnormality level of around 20 per cent would pertain at conception in man. This subject will be dealt with in more detail in the next section.

THE FREQUENCY OF ANEUPLOIDY AT CONCEPTION IN MAN

Let us now consider what is known about the initial incidence of chromosomal abnormality, and particularly aneuploidy, at conception in man and see how this compares with figures obtained for other mammalian species. Prior to the availability of techniques by which gametic genomes could be directly analysed, such calculations had to be based on figures for chromosomal anomaly among early spontaneous abortions.

Published estimates range from 8–10 per cent (Alberman and Creasy 1977; Kajii *et al.* 1978) through 20 per cent (Ford 1975) to 50 per cent (Boué *et al.* 1975), for the frequency of *all* chromosomally abnormal zygotes at conception in man, the wide variation in estimates arising largely because of the different methods used to calculate the figure. The lower estimates of Alberman and Creasy (1977) and Kajii *et al.* (1978) are based on the finding that about 15 per cent of all clinically recognized pregnancies abort spontaneously (mainly in the first trimester) and of these, some 50–60 per cent show karyotypic abnormalities, about 70 per cent of which are aneuploid (see Table 2.2) What is not taken into account, however, is the possibility that chromosome abnormalities among very early abortuses, i.e. those which do not produce a recognizable pregnancy, could be even more common than among abortuses screened in abortion surveys. Some human trisomics and virtually all autosomal monosomics are likely to be eliminated so early that the only sign of pregnancy would be, at most, a delayed menstrual period. Most autosomal monosomics in the mouse are eliminated before implantation (Gropp *et al.* 1974) and, in man, although sometimes found among early spontaneous abortions (e.g. Ohama and Kajii 1972), they are rarely recovered. These facts were taken into account by Ford (1975), who concluded that the proportion of human zygotes which are chromosomally abnormal at conception might be as high as 20 per cent. Ford's (1981) calculated current best estimate of the frequency of aneuploid human fetuses *at the time of clinical recognition of pregnancy* is 8 per cent.

The much higher figure of 50 per cent abnormalities at conception in man, suggested by Boué *et al.* (1975), is again based on the assumption of a 15 per cent spontaneous abortion rate for all clinically recognized pregnancies and a 60 per cent rate of chromosome abnormalities amongst them. Starting with 1000 conceptions, it is then postulated that 850 would go to term and 150 would be aborted; of the latter, 60 per cent or 100 zygotes would be chromosomally abnormal. From Boué's own cytogenetic survey of spontaneous abortions in the first twelve weeks of pregnancy (Boué *et al.* 1975), it was found that 15 per cent of all abnormalities were trisomic for chromosome 16 and a further 15 per cent were monosomic for the X chromosome. Boué *et al.* (1975) thus argued that if all the human chromosome pairs underwent nondisjunction with the same frequency as the X chromosome and chromosome 16, an equivalent number of zygotes monosomic and trisomic for all the other chromosome pairs would be produced. However, most of these would not produce clinically recognized pregnancies. If these assumptions were correct, there would then be $15 \times 23 = 345$ trisomics or a total of 690 anomalies resulting from nondisjunction alone. Combined with the other chromosome anomalies observed in abortions, notably polyploids, a figure for total chromosome abnormalities at conception in man quite close to that of 850 births at term, would thus be obtained. One problem with these calculations, however, as in the work described earlier for 'double spot' analysis in sperm, lies in the assumption that all human chromosome pairs undergo nondisjunction at the same rate. While this may be true, it seems unlikely. It is quite conceivable, as mentioned earlier, that chromosome pair No. 16 could show a greater propensity for nondisjunction owing to some special feature, e.g. a large heterochromatic block, possessed by it. The fact that trisomy 16 among abortuses shows little association with increasing maternal age (Hassold and Matsuyama 1979; Hassold *et al.* 1980*a*) implies that some special age-independent mechanism of induction is in fact operating in this common trisomic condition. Furthermore, X monosomy could arise through a mechanism other than nondisjunction, e.g. by lagging of a sex chromosome on the meiotic spindle or in an early cleavage division with its eventual exclusion from both daughter nuclei. In such cases of chromosome loss, there would be no corresponding trisomic product formed. X monosomics in man may, as in the mouse (Russell and Montgomery 1974), be far more common than could be predicted on the basis of an exclusively nondisjunctional origin (see Chapters 4 and 6). To use trisomy 16 and monosomy X as base levels for estimations of the overall frequency of nondisjunction in man and of the overall incidence of aneuploidy at conception could thus give a spuriously high figure. That real differences do exist in the rates of nondisjunction for particular human chromosome pairs is a possibility which Warburton *et al.* (1980*b*) have emphasized. These authors point out that if the rarer trisomics were likely to be lethal in an earlier stage of gestation, one might expect to find a correlation between the frequency with which a trisomy is detected and the mean gestation time at abortion. The available evidence does not support this (see Table 2.4). The rarer trisomics can clearly sometimes survive until after implantation, and are compatible with the development of embryonic structures. Furthermore, the trisomics which commonly survive to term (13, 18, 21, X) might be expected to be the least likely to be selected against very early in development. But these trisomics are among the most frequent in spontaneous abortions, although they are, nevertheless, much rarer than trisomy 16 (see Table 2.3). Small chromosomes, chromosomes associated with viable trisomics and acrocentric chromosomes are all found more frequently among abortuses, but which is the relevant characteristic is difficult to decide. Chromosome 16 has a large heterochromatic region and is a common trisomy, but chromosomes 1 and 9 have similar heterochromatic regions and are not common (Warburton *et al.* 1980*b*).

None the less, as Boué *et al.* (1975) have pointed out, their figure of 50 per cent is an estimate in good agreement with that deduced from some other observations in humans. The early work of Hertig *et al.* (1956) measured the extent of human errors of reproduction at a time before modern cytogenetic methods were developed. They studied the anatomical features of human zygotes 1–17 days old, recovered from the oviducts and uteri of 34 out of 211 hysterectomy patients. Eight were free-lying, one in the tube and seven in the uterine cavity, and 26 were implanted. Twenty-one of the 34 specimens were stated to be 'normal' and 13 were 'abnormal to one degree or another'. The authors found, in fact, that half of the free-lying ova and one-third of the implanted ova were morphologically abnormal. They were unable to predict, however, whether any of these pregnancies, from women of known fertility, would have terminated in abortion, although it seems likely that at least some would have done so. It is still not known, however, whether the morphological abnormalities observed were all or even partly attributable to chromosomal abnormality. For that, cytogenetic studies on the same type of material would be required and, if carried out, could provide results of great interest (Ford 1981).

One recent cytogenetic study which throws some light on the true incidence of chromosome abnormality in the early weeks of human gestation comes from Japan, where very young *induced* abortuses are available for study (Kajii *et al.* 1978). The figure for the frequency of chromosome abnormalities in embryos 21–34 days from conception (3–4 completed weeks) in these studies is put at 9.3 per cent (10 abnormals out of 108 karyotyped).

Whatever the figure, it would seem, however, that the total incidence of zygotic loss through chromosome abnormality may be very much higher in man, perhaps even of an order of magnitude, than is found in most other mammalian species, and the differences are even more marked if one considers only aneuploids. In most species which have been investigated, spontaneous levels of aneuploidy are much lower than in man (see Chapter 5).

Recent studies in the marmoset, *Callithrix jacchus*, suggest, however, that the loss of chromosomally abnormal fetuses in early pregnancy in that species may occur on a similar scale to that in man (Bobrow, M., personal communication), and it is thus tempting to suppose that there may be a fundamental biological difference between primates and other investigated species that accounts for the much higher levels of aneuploidy among their conceptuses.

ANEUPLOIDY IN HUMAN ABORTUSES AND STILLBIRTHS

The first chromosome anomalies described in man were associated with well known clinical syndromes compatible with live births. When, however, in 1960 the two lethal syndromes associated with trisomy D and trisomy E were found, it seemed logical to assume that additional severe chromosome anomalies would lead to disorders incompatible with life. This hypothesis received support from the description in 1961 of two abortuses, both of which had 69 chromosomes (triploidy) (Delhanty *et al.* 1961; Penrose and Delhanty 1961), and these discoveries heralded an era of cytogenetic study in abortuses. In the last twenty years of investigation, it has become abundantly clear that human spontaneous abortions are indeed a rich source of chromosomal abnormalities.

The estimated incidence of human spontaneous abortion, based on personal interviews of women in two American cities, is about 15 per cent of all pregnancies (Roth 1963; Warburton and Fraser 1964). This figure is the one most often quoted in studies concerned with cytogenetics of abortion, though it has been pointed out (Carr 1971), that it could be an underestimate of the real frequency because of the numbers of early fetal losses which might occur without the women even being aware they were pregnant. Incidence figures based on life tables therefore put the estimate as high as 24–29 per cent (Erhardt 1963; Bierman *et al.* 1965). A recent estimate, however, based on a method of diagnosing early pregnancy by rises in β HCG levels in urine (Williamson and Miller 1980), gave an apparent abortion rate of 13.7 per cent, a figure much more in line with that based on interviews with women. Pre-implantation losses cannot be detected by this assay but total post-implantation loss (before 20 weeks) was estimated to be at least 43 per cent (Miller *et al.* 1980).

The first serial cytogenetic study of abortions was carried out in 1963 (Carr 1963; Clendenin and Benirschke 1963) and since then a large number of reports have appeared in the literature involving chromosome analysis of both selected and unselected material (e.g. Boué and Boué 1973*a*; Boué and Thibault 1973; Creasy *et al.* 1976; Carr and Gedeon 1977; Hassold *et al.* 1978, 1980*b*; Warburton *et al.* 1980*b*). The actual frequency of chromosomal abnormality in spontaneous abortions has varied greatly from one survey to another, and this is due to a number of factors, not least of which is the small sample size of some of the studies, and the selection of specimens, both deliberate and unintentional. The most important criterion for selection is, however, gestational age of the fetus at expulsion, as calculated from the first day of the mother's last menstrual period. A glance at Fig. 2.1 shows how in one early but fairly typical abortion survey (Creasy *et al.* 1976), the incidence of chromosome anomalies fell as gestational age increased. In this particular survey, 50 per cent of abortions occurring in the first trimester of pregnancy (weeks 8–12) were chromosomally abnormal, while only 27 per cent of those occurring in the second trimester (weeks 13–24) were found to be so. The rate of chromosome anomaly in very early loss (i.e. before three weeks of gestation) is still not amenable to study because only specimens aborted after the first missed menstrual period can be collected. However, the striking observation has been made (Warburton *et al.* 1980*b*) that spontaneous abortions occurring

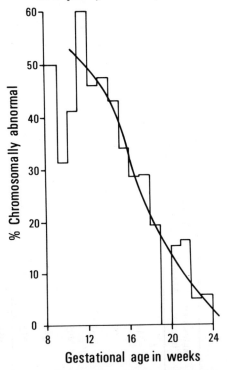

Fig. 2.1. Proportion of chromosomally abnormal abortuses by gestational age (weeks from LMP) at expulsion. The curve is an approximate best fit to the three-weekly running averages. (From Creasy *et al.* (1976), with permission.)

at 3–7 weeks gestation have a much lower rate of chromosome abnormality than do those occurring at later gestational ages (Table 2.1). Warburton *et al.* (1980*b*) consider various possible reasons for this, and suggest it may be the result of a trend to retain chromosomally abnormal specimens *in utero* for several weeks after implantation.

TABLE 2.1.   *Frequency of chromosome abnormalities in relation to gestational age of spontaneous abortions from surveys around the world (From Warburton* et al. *1980*b)

Weeks gestation from LMP	Honolulu (n = 672) % abnormal	Geneva (n = 446)	Hiroshima (n = 477)	London (n = 893)	New York (n = 876)
0–7	25.8	25.0	13.6	—	14.0
8–11	48.2	64.4	41.3	50.4	49.2
12–15	50.4	50.3	56.2	44.1	39.1
16–19	44.6	53.8	69.6	21.4	18.5
20+	27.0	0.0	40.0	6.6	11.1

TABLE 2.2.   *Frequency of different types of chromosome abnormalities found in six spontaneous abortion surveys*

Survey	45, X	3n	4n	Autosomal trisomy	Mosaicism	Structural abnormalities	Others	Total
Lauritsen et al. (1972)	12	3	4	14	0	1	0	34
Therkelsen et al. (1973)	39	14	10	66	4	6	0	139
Kajii et al. (1973)	12	10	5	51	0	3	1	82
Boué et al. (1975)	140	183	57	495	10	35	1	921
Creasy et al. (1976)	68	38	12	143	12	10	4	287
Hassold et al. (1980b)	112	70	33	212	12	20	4	463
Totals	383	318	121	981	38	75	10	1926
% of all abnormalities	19.9	16.5	6.3	50.9	2.0	3.9	0.5	

The single most common abnormality found among abortuses is the 45,X condition which comprises 20 per cent of all the abnormalities detected in six surveys quoted (Table 2.2). Altogether, during the prenatal period, it has been estimated that more than 95 per cent of all XO conceptuses abort (Simpson 1976), many during the second trimester (Singh and Carr 1966). Triploids are the second most frequent single group of anomalies, comprising 17 per cent, while autosomal trisomics, taken as a group, account for 50 per cent of the abnormalities. Structural and other abnormalities, including mosaics, sex-chromosome trisomics, and autosomal monosomics, account for only 6 per cent of cases. Taking the frequencies of X monosomy and the autosomal trisomics together, one can thus see from Table 2.2 that aneuploidy alone accounts for over 70 per cent

TABLE 2.3. *Trisomics among spontaneous abortuses. Results of eight surveys employing the use of banding techniques*

Trisomy	a	b	c	d	e	f	g	h	Total	%
1	—	—	—	—	—	—	—	—	0	0.0
2	4	6	1	2	1	2	21	16	53	5.6
3	1	1	—	3	—	—	2	1	8	0.8
4	2	1	2	1	2	1	11	4	24	2.5
5	—	—	—	—	—	—	1	—	1	0.1
6	1	—	—	—	—	—	2	—	3	0.3
7	6	—	2	4	1	3	17	10	43	4.5
8	4	3	3	2	—	4	13	6	35	3.7
9	1	3	3	2	1	—	11	6	27	2.8
10	4	1	2	1	—	1	5	4	18	1.9
11	1	—	—	—	—	—	1	—	2	0.2
12	2	—	—	—	—	1	5	—	8	0.8
13	5	3	1	1	4	1	24	15	54	5.7
14	10	3	4	2	2	5	9	5	40	4.2
15	11	4	11	1	2	1	25	14	69	7.3
16	39	35	21	14	10	15	110	64	308	32.4
17	—	—	—	—	1	—	4	2	7	0.7
18	7	6	6	4	—	—	15	10	48	5.1
19	—	—	—	—	—	1	—	—	1	0.1
20	1	2	—	—	—	1	14	8	26	2.7
21	12	10	3	3	1	4	34	12	79	8.3
22	19	11	4	4	6	1	38	13	96	10.1
Total	130	89	63	44	31	41	362	190	950	

a. T. Kajii. Personal communication to Warburton *et al.* (1980*b*).
b. Creasy *et al.* (1976).
c. J. E. Lauritsen. Personal communication to Warburton *et al.* (1980*b*).
d. H. Takahara. Personal communication to Warburton *et al.* (1980*b*).
e. F. J. Dill. Personal communication to Carr and Gedeon (1977).
f. Carr and Gedeon (1977).
g. Hassold *et al.* (1980*b*).
h. Warburton *et al.* (1980*b*).

TABLE 2.4 *Mean gestation time (days) for individual trisomics (Based on computerized data) (From Warburton et al. 1980b)*

Trisomy	n	Mean gestation	Standard deviation
2	15	81.2	22.4
4	3	74.3	3.4
7	7	88.6	21.8
8	4	104.3	29.6
9	5	74.6	3.4
10	3	96.0	1.4
13	9	82.9	21.4
14	4	100.3	30.4
15	10	96.5	28.7
16	48	82.7	17.4
17	1	87.0	—
18	8	98.6	53.7
20	5	88.6	16.1
21	9	107.3	25.9
22	8	84.5	12.2
X	4	82.7	12.1

of all the chromosomally abnormal spontaneous abortions, and makes by far the most important single contribution to all prenatal losses in man. Autosomal monosomics are rarely found, only four being reported in the six surveys. Yet our understanding of the origin of aneuploidy would lead us to expect monosomy arising by nondisjunction to be equally frequent as the corresponding trisomy. In fact, the total amount of monosomy, considering that arising also by chromosome loss, should exceed trisomy. The general belief, as stated earlier, is that the vast majority of autosomal monosomic conceptuses in man abort in the very early stages of pregnancy without producing recognizable pregnancies.

By means of banding for chromosome identification (Paris Conference 1971) it is possible to be precise in the identification of the various autosomal trisomics contributing to the abortions, and the most recent surveys have adopted these techniques (Table 2.3). The most common single trisomy is an extra chromosome 16, a fact established even before banding was introduced, and subsequently confirmed by all authors using the banding methods. Trisomy 16 accounts for one-third of all identified autosomal trisomic abortuses. No example of a trisomy 1 has yet been reported and trisomies 5, 6, 11, 12, 17, and 19 are extremely rare. After trisomy 16, trisomy 22 and trisomy 21 are the next most common aneuploids. The mean gestation times for individual trisomics are shown in Table 2.4.

Chromosome abnormalities occur in about 5 per cent of stillborn infants (Machin and Crolla 1974; Kuleshov 1976; Alberman and Creasy 1977;

Sutherland *et al.* 1978). In the United Kingdom viability of the fetus is conventionally assumed to be attained by the 28th week after the first day of the mother's last menstrual period, and the delivery of a dead fetus after this stage of pregnancy is defined as a stillbirth (Alberman and Creasy 1977). The abnormalities found are similar in type to those found in newborns but are about ten times more common. The highest prevalence is among macerated stillbirths, being about double that seen in fresh stillbirths and early neonatal deaths. The abnormalities reported are predominantly trisomy 13, 18, and 21; sex-chromosome aneuploidy; and unbalanced translocations. The perinatal mortality of trisomies 13 and 18 is much higher than that of trisomy 21, being over 90 per cent for the two former and only about 10 per cent for the latter (Machin and Crolla 1974; Alberman and Creasy 1977). Monosomy X has a higher death rate in the perinatal period than other sex chromosome aneuploids (Robinson 1974; Alberman and Creasy 1977).

It is not at all clear why many of the abnormal karyotypes found in man can show such a great range of phenotype, or why, although they have such a high mortality during the early stages of embryogenesis, a small proportion can survive so long into postnatal life. A plausible hypothesis would be that the phenotype of a trisomy is determined by the genetic constitution of the three chromosomes forming the trisomy. Those aneuploids that do survive will be the subject of discussion in the next part of the chapter. The *primary* incidence of aneuploidy is largely of academic interest. The genetic burden is determined by the frequency at birth and beyond.

ANEUPLOIDY IN HUMAN LIVEBORNS

From chromosome surveys of consecutive liveborn babies, we now have a good knowledge of the incidence of aneuploidy at birth in man. Such surveys have been carried out in at least ten laboratories throughout the world, more recently using banding techniques, and from the pooled data, the chromosome constitution of 64 887 babies, of whom 22 973 are females and 41 914 are males is known (Table 2.5). The excess of males is the result of one survey (Walzer and Gerald 1977) screening only males, and another (Jacobs *et al.* 1974), initially screening only males. Similar large surveys have also been carried out to estimate the incidence of sex chromosome aneuploidy among human newborns by using buccal smears but the results have been generally in good agreement with those based on a full chromosome analysis in respect both to the number of major sex chromosomal aneuploids and to the number of mosaics recognized (Jacobs 1979).

The chromosome data presented in Table 2.6 show that 88/41 914 or one in every 476 liveborn males and 28/22 973 or one in every 820 liveborn

TABLE 2.5.   *Frequency of aneuploidy in the human newborn population (Mosaics excluded) Results of ten independent chromosome surveys*

Survey	Number of individuals karyotyped			Sex aneuploids				Autosomal aneuploids			Total aneuploids	%
	♀	♂	Total	XYY	XXY	XXX	XO	+D	+E	+G		
1 London (Canada) 1969	1015	1066	2081	4	1	0	0	0	0	2	7	0.34
2 New Haven (USA) 1970	2181	2184	4365	3	4	3	1	1	1	3	16	0.37
3 Edinburgh (UK) 1974	3831	7849	11 680	10	9	5	0	0	2	17	43	0.37
4 Moscow (USSR) 1974	1197	1303	2500	0	1	0	0	0	0	4	5	0.20
5 Winnipeg (Canada) 1975	6763	7176	13 939	4	6	5	0	1	2	14	32	0.23
6 Arhus (Denmark) 1975	5387	5761	11 148	5	8	8	3	1	1	16	42	0.38
7 Jerusalem (Israel) 1975	241	259	500	0	1	0	0	0	0	0	0	0.00
8 Calgary (Canada) 1976	437	493	930	2	0	1	0	0	0	0	3	0.32
9 Boston (USA) 1977	—*	13 751	13 751	11	9	—*	—*	0	0	19	39	0.28
10 Edinburgh (UK) 1980	1921	2072	3993	4	6	3	0	0	1	3	17	0.37
Totals	22 973	41 914	64 887	43	45	24	4	3	7	78	204	0.31

* Females not studied.

References

1. Sergovich, F. *et al.* (1969). *New Engl. J. Med.* **280**, 851–5.
2. Lubs, H. A. and Ruddle, F. H. (1970). *Science, NY* **169**, 495–7.
3. Jacobs, P. A. *et al.* (1974). *Ann. hum. Genet.* **37**, 359–76.
4. Bochkov, N. P. *et al.* (1974). *Humangenetik* **22**, 139–52.
5. Hamerton, J. *et al.* (1975). *Clin. Genet.* **8**, 223–43.
6. Nielsen, J and Sillesen, I. (1975). *Humangenetik* **30**, 1–12.
7. Cohen, M. M. *et al.* (1975). *Israel J. med. Sci.* **11**, 969–77.
8. Lin, C. C. *et al.* (1976). *Hum. Genet.* **31**, 315–28.
9. Walzer, S. and Gerald, P. S. (1977). In *Population cytogenetics. Studies in humans* (ed. E. B. Hook and I. H. Porter) pp. 45–61. Academic Press, New York.
10. Buckton, K. E. *et al.* (1980). *Ann. hum. Genet.* **43**, 227–39.

TABLE 2.6. *Overall frequencies of sex chromosome and autosomal aneuploids in the human newborn population (Mosaics excluded) Data obtained from ten independent chromosome surveys (See Table 2.5 for references)*

(a)	Sex aneuploids	%	Frequency
♂	XXY	0.11	1 in 931 liveborn males
	XYY	0.10	1 in 974 liveborn males
	All	0.21	1 in 476 liveborn males
♀	XXX	0.10	1 in 957 liveborn females
	XO	0.02	1 in 5743 liveborn females
	All	0.12	1 in 820 liveborn females
	Overall incidence	0.18% or	1 in 559 live births

(b)	Autosomal aneuploids	%	Frequency
	+13	0.005	1 in 21 629 live births
	+18	0.01	1 in 9 269 live births
	+21	0.12	1 in 831 live births
	Overall incidence	0.14% or 1 in 737 live births	
	Total for **all** aneuploids	0.31% or 1 in 318 live births	

females has a sex chromosomal aneuploidy, while 88/64 887 or one in every 737 liveborns has an autosomal aneuploidy. The overall frequency of aneuploidy is 204/64 887 or one in every 318 livebirths. This compares with a frequency of one in every 425 liveborns carrying a structural chromosome abnormality (Nielsen and Sillesen 1975).

There is no evidence to suggest that any of these numerical abnormalities (with the possible exception of some X monosomics which might arise by sex chromosome loss following anaphase lagging at meiosis or early cleavage), are anything but primary whole chromosome trisomics or monosomics arising *de novo* as errors of segregation at meiosis in gametogenesis of one or other of the parents (Jacobs 1972), although theoretically, some could arise by very early mitotic nondisjunction. Partial or 'segmental' aneuploids are however excluded. Mosaics do of course exist (about 14 per cent of all sex chromosomal aneuploids identified at birth, for example, were found to be mosaic, with mosaics being twice as common in females as in males (Jacobs 1979)). They usually show some of the features of the particular aneuploid condition, but in a milder form than the manifestation seen in the full trisomic or monosomic (Hamerton 1971*a*). Mosaics have, however, been excluded from the data given in Tables 2.5 and 2.6.

The figures for aneuploidy among liveborns is of course dramatically lower than estimates given for the incidence at conception in man and this illustrates strikingly the role of spontaneous abortion and perinatal mortality in eliminating most aneuploid fetuses along with many other chromosomally abnormal conceptuses. Such a process of elimination can be viewed as a natural function responsible for maintaining genetic stability and ensuring the quality of surviving fetuses. The human aneuploids which do reach full term are thus the tip of a huge iceberg.

Three autosomal trisomics survive more commonly to term, namely D13, E18, and G21, and are sufficiently common at birth to have well-described syndromes, those respectively of Patau, Edwards, and Down. Down syndrome was first described in 1866 by Langdon Down, Physician to the London Hospital and the Earlswood Asylum, but the chromosomal nature of the disorder was not demonstrated, by Lejeune *et al.*, until 1959. The rarer Edwards and Patau syndromes were described as trisomics in 1960 by Edwards *et al.* (1960) and Patau *et al.* (1960) respectively. From the newborn survey data presented in Table 2.6, trisomy D is found in about 1 in 21 000 livebirths, trisomy E in about 1 in 10 000 livebirths and trisomy G in about 1 in 800 livebirths. These figures are only averages hiding much heterogeneity, for the incidence of all three aneuploid conditions increases with the age of the mother (Taylor 1968; Smith and Berg 1976) (see Chapter 4).

The much higher incidence of the trisomy 21 condition at birth than of trisomies 13 and 18 could be explained either on the basis of a proportionately greater prenatal loss of D and E trisomic fetuses or on an increased initial frequency of nondisjunction in the G group compared with the D and E groups. From data obtained in abortion surveys which have employed the use of banding techniques there appears, however, to be little difference in the frequency of the trisomy 13, 18, and 21 conditions among abortuses (see Table 2.3) and the second explanation would thus seem to be the more likely. However, it is also possible that greater loss of D- and E-group compared with G-group trisomics occurs among very early embryos which go undetected in the abortion surveys.

Of all the autosomal aneuploids conceived, however, the prenatal loss of D- and E-trisomics is over 90 per cent (Alberman and Creasy 1977) and of G-trisomics 65–80 per cent (Creasy and Crolla 1974; Kajii *et al.* 1973; Boué *et al.* 1981). Furthermore, of those which do survive to term, few D- and E-trisomics survive beyond the first six months of life (Weber 1967; Taylor 1968), many dying in the perinatal period with multiple malformations (Machin and Crolla 1974; Alberman and Creasy 1977). Some G-trisomic individuals, particularly those with cardiac defects, also die in their first year of life, but others survive longer into adult life, albeit with a shorter life expectancy than the normal population (Smith and Berg 1976). Improved chances of survival of Down syndrome individuals in recent years have followed on health measures and higher standards of living (Carter 1958). Their respiratory infections now yield to antibiotics, their low resistance to other infections is supported by immunization, and their anomalies are corrected by surgery. The severe mental defect is, however, irreversible. For every 100 people born with Down syndrome, Hagard and Carter (1976) have calculated, from the data of two comprehensive surveys, that 76 would still be surviving at one year of age, 69 at five years, 27 at forty-five years and 11 at sixty years.

A number of less common autosomal aneuploids occasionally also survive to term in man. Among these are trisomy 8 (see Riccardi 1977, for review), trisomy 22 (e.g. Uchida *et al.* 1968; Bass *et al.* 1973; Penchaszadeh and Coco 1975), trisomy 9 (Feingold and Atkins 1973; Sutherland *et al.* 1976; Seabright *et al.* 1976), and trisomy 7 (Yunis *et al.* 1980). Most liveborn subjects with trisomy 8 exhibit various manifestations of the 'Warkany' syndrome, characterized by mental retardation, bone and joint anomalies, and various other physical defects. The mental retardation is usually mild, however, compared to that seen for example in trisomy 21. Babies born with the trisomy 22 condition suffer severe multiple congenital malformations and mental retardation (Penchaszadeh and Coco 1975). Among the reported cases, there have been three pairs of sibs with the condition (Uchida *et al.* 1968; Gustavson *et al.* 1972; Zackai *et al.* 1973), a proportion that seems higher than is usually seen in nondisjunctional chromosome disorders. In one case, however (Uchida *et al.* 1968), the mother of the two sibs was herself a trisomy 22 mosaic and in another case (Zackai *et al.* 1973), the mother of the affected sibs had a terminal deletion in the long arm of a No. 22 chromosome, a factor which the authors suggested may have predisposed this chromosome pair to undergo nondisjunction. The trisomy 9 case reported by Feingold and Atkins (1973) was a male who survived for 28 days; that of Seabright *et al.* (1976) was a female who survived for 16 hours. The trisomy 7 case reported by Yunis *et al.* (1980) was a female who lived for two days. All were severely affected with multiple malformations at birth.

Among the more common sex chromosomal aneuploids, those which can survive to term are the 45,X and 47,XXX conditions in females and the 47,XXY and 47,XYY conditions in males. X monosomics occur with a frequency of 1 in 5743 among liveborn females and are much rarer than 47,XXX individuals, who are found with a frequency of about 1 in 1000 liveborn females (Table 2.6). The much lower incidence of XO females among the newborn is explained by great loss of 45,X fetuses during gestation. In the abortion surveys, monosomy X alone accounts for nearly 20 per cent of all the chromosome abnormalities found among spontaneous abortuses (see Table 2.2). It is possible also that additional X monosomics are missed in the abortion surveys because they are eliminated before producing a recognizable pregnancy. Some also die perinatally (Machin and Crolla 1974; Robinson 1974; Alberman and Creasy 1977). As stated before, it has been estimated that more than 95 per cent of all human XO conceptuses are lost at some time or other in the prenatal period (Simpson 1976).

Those few X monosomics which do survive to term are typically associated with Turner syndrome (Turner 1938). In women with this syndrome, the ovaries are normally replaced by fibrous tissue with a histological structure of ovarian stroma but devoid of follicles. The condition is characterized by primary amenorrhoea, infantile genitalia, dwarf stature, and a variable array of congenital abnormalities including a short webbed neck, low hair-line, and cardiovascular and renal defects (Lindsten 1963). Life expectancy in Turner women is less than for women in the general population (Price, W.H., personal communication). The malformations are, however, much less severe than those seen in most of the viable autosomal trisomic conditions, and it is remarkable, in view of the high mortality rate of XO conceptuses in early embryogenesis, that a small proportion should survive so long into adult life.

The 47,XXX chromosome condition was first described by Jacobs *et al.* (1959) in two amenorrhoeic women. A considerable number of 'triple-X' females have since been reported, and the clinical findings have been variable and the phenotype unremarkable (Barr *et al.* 1969). In fact there is no clear-cut syndrome associated with this particular chromosome constitution, though there is a mild to moderate lowering of intelligence (Ratcliffe *et al.* 1979). The 47,XXX condition, like 47,XXY and 47,XYY, is extremely rare in abortion material (Jacobs 1979), but is sometimes found among perinatal deaths (Machin and Crolla 1974; Alberman and Creasy 1977). All three of these sex aneuploids are generally compatible with life. Life expectancy among XXY males appears to be normal (Price, W.H., personal communication); data on longevity among XXX and XYY subjects are not available.

47,XXY trisomics are typically associated with Klinefelter syndrome (Klinefelter 1942) characterized by aspermatogenesis, tall eunuchoid proportions, occasional gynaecomastia, and various other phenotypic abnormalities. With increasing numbers of clinical reports of these patients, it is clear that the picture is very varied, depending for its severity on the ascertainment source of the patients. A small but significant reduction of intelligence is described for the group as a whole, but patients have been described with above-average intelligence and working in professional occupations such as the law and medicine. Most show imperfect masculinization at puberty, with little facial hair, absence of body hair, and a female distribution of pubic hair. The testes are very small but the penis usually is of normal size. The condition contributes significantly to infertility in the male, the frequency of such individuals among males attending infertility clinics being 10 per 1000 compared with only 1 per 1000 in the newborn male population (Chandley 1979).

The phenotype associated with the 47,XYY condition is often even more symptomless, the early cases having been referred because of hypogonadism, abnormal external genitalia, Marfan syndrome, occasional cryptorchidism, and various other non-specific features (Hamerton 1971*b*). Subsequently, however, a few individuals have been ascertained at both paediatric clinics and adult psychiatric clinics because of behavioural problems and abnormal height. Interest in the effect of the two Y chromosomes was aroused initially by the finding of a high proportion of 48,XXYY males in institutions for the criminally insane (Casey *et al*. 1966, 1968; Jacobs *et al*. 1965, 1968). Jacobs *et al*. (1965) studied the chromosomes of 315 male patients in a Scottish maximum security hospital and among these, found nine 47,XYY males, six of whom were over 180 cm tall. From this and other studies, it became clear that most of these males were characterized by increased height, and in more recent surveys, this has been one of the main selection criteria. The physical characteristics of 47,XYY males ascertained in surveys show that while the majority are of slightly below normal intelligence, a proportion are average. The vast majority have a normal male habitus with normal sexual development, though there is a suggestion of an increased risk of spermatogenic impairment within the group (Skakkebaek *et al*. 1973*b*).

Although more than fifteen years have passed since the first explosion of interest in the XYY complement, following reports that the double representation of the Y chromosome in the karyotype was associated with criminal behaviour (Jacobs *et al*. 1965; Casey *et al*. 1966), the consequences of an XYY sex chromosome complement remain in doubt. The question of whether an XYY chromosome complement commonly, occasionally or only rarely predisposes to antisocial behaviour remains to be clarified by the outcome of prospective studies on XYY children. Even some who do find evidence of an increased probability of criminality question the popular belief that it is aggressive behaviour that brings the XYY individual into the courts. They believe rather that it is the lowered intelligence and other neurological alterations associated with the XYY condition that are implicated as the important mediating variables (Witkin *et al*. 1976).

From the foregoing, it is evident that autosomal and sex-chromosomal aneuploids represent a very significant proportion of all chromosomally abnormal liveborn individuals. The phenotypic abnormalities associated with each of the viable monosomic or trisomic conditions all give cause for concern. For many of the sex chromosomal aneuploids without obvious diagnostic physical features at birth, there is an uncertain prognosis, and convincing evidence that at least some are at an increased risk from a variety of problems (emotional, physical, and intellectual) in later life.

Current prospective studies of such children are therefore very valuable in providing the foundations of appropriate care for affected children and in giving guidance to their parents (Robinson 1979).

FERTILITY AND REPRODUCTION IN LIVEBORN HUMAN ANEUPLOIDS

In view of the possibility that aneuploidy can be generated by secondary nondisjunction (see p. 49) at meiosis in a fertile aneuploid individual at reproduction, it seems appropriate to consider with what frequency secondary aneuploidy arises in man. This section of the chapter is therefore devoted to a consideration of the fertility and reproductive potential of viable aneuploid individuals.

In the case of the sex chromosome aneuploids, we are mainly dealing with the XXY and XYY males and the XXX and XO females.

As is well known, the testes of XXY males with Klinefelter syndrome are small, hyalinized, and devoid of germ cells and the seminal analysis usually reveals azoospermia. Fertile tubules have, however, occasionally been found in the testes of such men (Skakkebaek *et al.* 1969) and spermatozoa reported in their ejaculate (Foss and Lewis 1971), even in cases where no XY/XXY mosaicism was suspected. Paternity has even been claimed in a few cases (Schiavi *et al.* 1978). Burgoyne (1978) believes that spermatogenesis in a non-mosaic XXY male could result from localized repopulation by XY germ cells which had originated from XXY germ cells by disjunctional accidents or X chromosome loss. The generally observed sterility of XXY males both in man and many other investigated mammalian species has led to the postulate that the presence of two X chromosomes in a testicular germ cell results in its perinatal death (Lyon 1974; Burgoyne 1978).

At birth, the testes of a group of 18 non-mosaic XXY babies were found by Ratcliffe *et al.* (1979) to be of normal size and consistency, and in one four-week-old infant studied by these authors, a normal histological appearance was found on testicular biopsy. By the age of six months, however, the testes in most boys had decreased in size and become unusually soft in consistency. Normal testicular growth up to adolescence was, however, recorded in two boys (Ratcliffe *et al.* 1982).

The histological changes which occur in the XXY testis are mainly prepubertal. Mikamo *et al.* (1968) found the number of spermatogonia in three boys aged three, four, and twelve months to be only 24, 18, and 0.1 per cent of control values. In boys aged 7–12 years, Ferguson-Smith (1959) reported two types of tubule in the testis; most lacked spermatogonia but other occasional fertile tubules showed spermatogenesis. Evidence of spermatogenic activity has, however, been found in an occasional pubertal subject (Bunge and Bradbury 1956; Ferguson-Smith and Munro 1958).

The testes of newborn XYY babies are usually normal (Ratcliffe *et al.* 1979) and there are no differences in testicular size between XYY men and XY controls (Schiavi *et al.* 1978). The histological appearance of the adult XYY testis, even among individuals not ascertained through infertility, has, however, been found to show considerable variability, some individuals having normal spermatogenic activity, others a mild to severe degree of maturation impairment (Skakkebaek *et al.* 1973*b*). The percentage of affected tubules in the biopsies has ranged from 0 to 87 per cent. However, the frequency of the XYY karyotype among subfertile males attending infertility clinics is no higher than among controls in the newborn male population (Chandley 1979).

At meiosis in most XYY men who have been investigated, the spermatocytes at diakinesis/metaphase I in air-dried preparations have rarely shown the presence of the extra Y chromosome, although some XYY spermatogonial metaphases have usually been found (Thompson *et al.* 1967; Evans *et al.* 1970; Tettenborn *et al.* 1970; Chandley *et al.* 1976*b*). An exceptional case described by Hultén and Pearson (1971) showed 45 per cent of primary spermatocytes at diakinesis containing what the authors interpreted as a YY bivalent, when fluorescence staining was applied. Moreover, electron microscopical analyses of serially sectioned spermatocytes from human XYY males, carried out by Berthelsen *et al.* (1981), have indicated that, in three out of four patients, a significant proportion of germinal cells retained the extra Y chromosome until at least the pachytene stage. The authors believe that earlier failures to detect the extra Y chromosome were mainly caused by the different techniques employed. However, the absence of the extra Y chromosome from the testes of adult XYY human males has been explained by Evans *et al.* (1970) as perhaps being due to accidental Y-loss from a primitive germ cell or spermatogonium and the subsequent proliferative advantage of the XY germ cell so produced. Alternatively, since XYY spermatogonia are usually found in XYY men, there may be *directed* loss of the extra Y at a late stage in maturation of the spermatogonium or spermatocyte (Burgoyne 1979). Whatever the mechanism, the risk of an XYY male passing his extra Y chromosome to his sons appears negligible. Among fertile XYY men whose offspring have been karyotyped, normal XY sons and XX daughters have been reported (Thompson *et al.* 1967; Court Brown 1968; Lisker *et al.* 1968).

Turning now to the female sex-chromosome aneuploids, let us consider the XO female with Turner syndrome. This condition in adult women is normally characterized by primary amenorrhoea, infantile genitalia, and failure of secondary sexual development. The ovaries are absent or rudimentary and replaced by fibrous tissue with a histological structure of ovarian stroma but devoid of follicles. They are usually referred to as 'streak gonads' (Hamerton 1971*b*). The XO ovary during fetal and prepubertal development is, however, variable in appearance. Up to three months gestation, it contains germ cells in normal numbers (Singh and Carr 1966; Carr *et al.* 1968) and even at birth, some oocytes may still be present (Gordon and O'Neill 1969). However, the ovaries in most XO subjects are totally devoid of oocytes at birth, the gonad consisting only of a thin band of tissue as in the adult streak gonad (Carr *et al.* 1968; Peters *et al.* 1981). Nevertheless, menstruation and pregnancy have been recorded in a few apparently non-mosaic XO women (King *et al.* 1978; Lajborek-Czyz 1976; Nakashima and Robinson 1971).

The cause of the early loss of germ cells in XO individuals is not clear, but X-dosage deficiency brought about by the absence of one normally active X chromosome in the oocyte has been suggested as the possible genetic basis (Lyon 1974; Burgoyne 1978). Jirásek (1977), on the other hand, has observed that there is incomplete enclosure of the oocytes by the developing follicles in XO fetuses and suggests that this could be the cause of the subsequent oocyte loss. As Burgoyne (1978) has pointed out however, these two viewpoints are not mutually exclusive: X-deficient oocytes may induce poor follicular development, and this in turn may accelerate the processes leading to oocyte death. By contrast, XO mice are fertile. Nevertheless, they are subject to considerable reproductive impairment and reproductive lifespan is markedly reduced owing to premature exhaustion of the supply of oocytes (Lyon and Hawker 1973). Lyon and Hawker (1973) and Burgoyne and Biggers (1976) have argued that the important difference between XO mice and XO women may simply be one of time scale, XO mice reaching puberty before X-deficiency effects in the oocyte become severe, XO women reaching puberty after all oocytes have degenerated. The available data on ovarian morphology and fertility for other mammalian XOs is consistent with this hypothesis; XOs from species with a short generation time being fertile, whereas those from species with a longer generation time are sterile (Burgoyne 1978).

Triple-X women are phenotypically unremarkable and usually fertile (Barr *et al.* 1969); most menstruate normally and many have children (Stewart and Sanderson 1960). However, XXX women with histories of irregular menstruation have been described, and when it was possible to examine such individuals at operation, their ovaries contained few follicles and were comparable to those seen in women at or near the menopause (Johnston *et al.* 1961). The children of fertile XXX women are usually chromosomally normal, suggesting that some mechanism operates during meiosis to eliminate the extra X chromosome from the egg. Hamerton (1971*b*) has suggested that some form of directed segregation may occur during oogenesis so that the disomic chromosome complement is regularly driven into one of the polar bodies.

For the viable autosomal aneuploids, a good deal of information is now available on the fertility and reproduction of trisomy 21 individuals.

Children with Down syndrome often show poorly developed sexual characteristics and late sexual development (Smith and Berg 1976) and in a survey of the ovaries of girls at different ages, all have been found to be abnormal, containing fewer small follicles than the ovaries of normal children (Højager *et al.* 1978). Furthermore, the atresia rate is altered; the number of follicles declines rapidly after the age of three years. In many ovaries, follicular growth is absent and no antral follicles formed, while in others, an abnormally low number of antral follicles is present. None of the ovaries examined show the active follicular growth typically seen in the ovaries of normal girls (Peters *et al.* 1981). The ovaries of trisomy 18 babies have also been studied by Alexiou *et al.* (1971) and Russell and Altschuler (1975). They too show severe reductions or even total absence of oocytes and follicles.

According to Smith and Berg (1976), at least 23 examples of fully affected trisomy 21 adult females who have given birth to children are known. Their 24 progeny have consisted of approximately equal numbers of chromosomally normal and G-trisomic individuals (the Mendelian expectation).

There appears to be no report of a fully affected Down syndrome male fathering a child (Smith and Berg 1976). Incomplete descent of the testes is frequent and, according to Benda (1969), occurs in approximately 50 per cent of all cases. The testes, when normally descended, are unusually small, and the testicular histology shows generally a degree of impaired spermatogenesis, which is often severe (Skakkebaek *et al.* 1973*a*), but spermatozoa are sometimes found in fairly good numbers in those Down males who are able to produce an ejaculate (Stearns *et al.* 1960). The mean counts, however, are usually lower than the average levels for normal fertile men. The inability of Down males to reproduce appears to be related as much to their sexual impotence as to their inability to produce gametes in sufficient number.

Meiosis has been studied in a number of male Down individuals (e.g. Sasaki 1965; Finch *et al.* 1966; Hultén and Lindsten 1970; Kjessler and de la Chapelle 1971) and G trivalents, two G bivalents plus a G univalent, and two G bivalents alone have been recorded. In the latter cells, which appear, however, to constitute a minority, the extra chromosome 21 is missing from the complement.

In view of the fact that so few other viable autosomal aneuploids survive to reproductive age, little can be said of their fertility. A recent study of a trisomy 8 mosaic male with normal intelligence and near-normal phenotype has, however, been made (Chandley *et al.* 1980). He showed infertility due to severe oligospermia. The testicular histology indicated a severe degree of maturation impairment in many of the germ cells and the mean sperm count was less than 1 million per ml. Meiotic studies showed that the extra No. 8 chromosome was absent from the germ line.

From the foregoing, the conclusion can be reached that natural selection severely limits the opportunities for reproduction in liveborn aneuploid individuals, and thus removes from the population a great potential source of further aneuploidy. The failure of germ-cell function in the XXY, XO and possibly XYY conditions and the possible elimination of the extra chromosomal element in the XYY and XXX subjects reported, all combine to prevent aneuploidy being passed on from one generation to the next. Only in the case of the female with Down syndrome does it seem that the nondisjunctional error is likely to be repeated from parent to offspring. Confinement within an institution for the majority of such women ensures that they too are excluded from the pool of reproductively active members of society.

3. The origins and causes of aneuploidy in experimental organisms

In the preceding chapter we considered the impact of meiotic aneuploidy on society, and the extent to which it is a problem. In Chapter 4 the factors which might be responsible for bringing about this high genetic load in man will be considered in detail, but first it is important to understand the types of cellular defect which generate aneuploidy.

To help in achieving this objective, data will be considered from a range of taxonomically divergent organisms, and later, the question of whether any or all of the proposed mechanisms could operate in man, will be considered.

To the affected individual the precise origin of the chromosomal defect might seem unimportant. None the less, the exact cause of aneuploidy may be of more than academic interest. Currently there is widespread concern that chemical compounds may exist which can induce aneuploidy as their sole genetic end-point (see Chapter 7). If this concern is real, the efficiency with which such trisomigens are detected will be increased by a thorough knowledge of the causes of aneuploidy. Not all testing methods detect aneuploidy from each source, and some agents may specifically affect only part of the cellular machinery, leaving other parts undisturbed. It is important, therefore, to recognize the limitations (if any) of the various testing methods.

Although the definition of aneuploidy is straightforward enough, several different kinds of trisomy have been recognized which have different underlying causes:

Primary trisomy—In this case the extra chromosome present in the cell is a structurally normal member of the complement.

Secondary trisomy—Here the extra chromosome is an isochromosome, i.e. one in which both arms are genetically and structurally identical. Isochromosomes are thought to arise by misdivision of the centromere, with the plane of division being across the chromosome rather than along it. Secondary trisomics are identified by the presence of rings of three chromosomes at meiosis.

Tertiary trisomy—Here the extra chromosome results from chromosome breakage and translocation, so that the extra chromosome possesses arms from two non-homologous chromosomes.

The meiotic configurations formed at metaphase in the three types of trisomic are illustrated in Fig. 3.1.

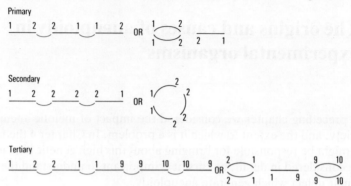

Fig. 3.1. Meiotic configurations found at metaphase in primary, secondary, and tertiary trisomics. The numerals identify chromosome arms. In the secondary trisomic a chromosome with identical arms is present. In the tertiary trisomic a chromosome with a novel combination of non-identical arms is present. (From Swanson, Merz, and Young (1981), with permission.)

DEFECTS LEADING TO ANEUPLOIDY

In this, and subsequent sections of the chapter, consideration will be given to the mechanisms relating to only the first of these three types of aneuploidy, i.e. primary trisomy, which generates extra or missing whole chromosomes. The literature contains a number of postulated defects (and experimental evidence for their occurrence), which could lead to primary trisomy in addition to the classical pathways of 'nondisjunction' and 'chromosome loss'.

Nonconjunction

Aneuploidy may arise from an early defect in meiosis in which homologous chromosomes fail to pair or fail to maintain a paired state. Sturtevant and Beadle (1939) used the term 'nonconjunction' for this defect. Following pairing failure, aneuploid products may be generated if univalent chromosomes segregate randomly at anaphase I. Sturtevant and Beadle believed that the majority of aneuploid meiotic products were formed in this way. In some organisms there is evidence that, if homologous chromosomes are unpaired, their subsequent segregation may be influenced by association with non-homologous chromosomes (Grell 1959; Novitski and Puro 1978).

Nondisjunction

Bridges (1913) accounted for sex-chromosome aneuploids in *Drosophila* by postulating nondisjunction in which 'the two X chromosomes fail to

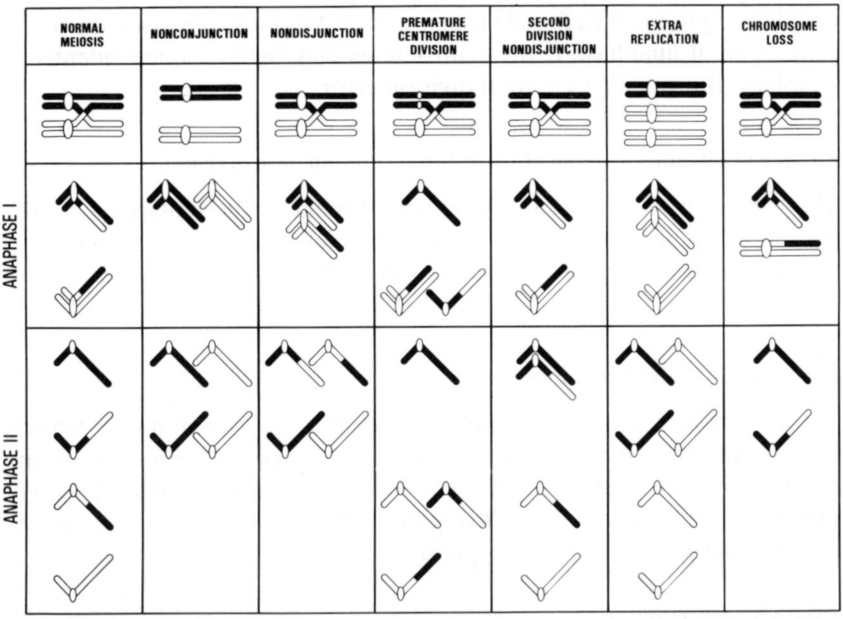

NORMAL MEIOSIS	NONCONJUNCTION	NONDISJUNCTION	PREMATURE CENTROMERE DIVISION	SECOND DIVISION NONDISJUNCTION	EXTRA REPLICATION	CHROMOSOME LOSS

Fig. 3.2. Schematic representation of various cellular defects giving rise to meiotic aneuploidy.

disjoin from each other. In consequence both remain in the egg or both pass out into the polar body'. He believed that failure of separation came about through entanglement of homologues. Another possibility is that nondisjunction could result from malorientation of microtubules so that both components of a normally synapsed bivalent move to the same pole.

Defective centromere division

Aneuploidy may also arise from an error of centromere segregation at the first meiotic division. Should the centromere of one of the homologous chromosomes in a bivalent divide prematurely during the first meiotic division then, after anaphase I, both of the daughter cells will contain a single unpaired chromatid. Assuming its homologue does not show the same defect, the resulting meiotic products will be aneuploid (see Fig. 3.2). Polani and Jagiello (1976) have called the process giving rise to single chromatids 'pre-division', whilst Hansmann and El-Nahass (1979) have used the term 'pre-segregation'. Threlkeld and Stoltz (1970) used the descriptive term 'precocious centromere division' for this defect. Work with *Drosophila* mutants has shown that defective centromere division may lead to an error at either the first or second meiotic division. That is, the defective centromere may result in the segregation of sister chromatids at anaphase I producing a first division error or, following normal reductional separation at anaphase I, the centromeres may behave independently at anaphase II producing a second division error.

Failure of chromatid separation at the second meiotic division

This error may come about either through malorientation of the microtubules, or through a centromere defect preventing chromatid separation. In either case the result will be nondisjunction at the second division.

Extra replication of chromosomes

The possibility exists that aneuploidy may arise, not from a cell division defect, but from an error of chromosome replication at some time during meiosis, so that an extra copy of a chromosome is generated. This will result in the production of hyperploid meiotic products without the concomitant production of hypoploids.

Chromosome loss

There are several possible ways in which a chromosome may fail to be incorporated into a daughter cell. For example, spindle fibres may fail to attach to a chromosome which consequently remains behind on the metaphase plate. Errors such as this are grouped under the general term 'chromosome loss'.

GENETIC METHODS OF DETECTING ANEUPLOIDY

The essential feature of genetic methods for detecting aneuploidy is that the aneuploid condition is inferred indirectly by the use of appropriate mutations, which enable the inheritance of marked chromosomes to be followed. This contrasts with cytogenetic methods where the aneuploid state is observed directly.

   In some systems the aneuploidy has a phenotypic consequence and the genetic markers serve simply to indicate the source or origin of the extra chromosome. In other systems the mutations themselves play an essential part in revealing the presence of the aneuploidy. As we shall see, no one genetic method allows the unambiguous determination of the underlying cause of the aneuploidy to be made in every case.

Aneuploid detection in *Drosophila*

Bridges (1916) was the first systematically to study the abnormal inheritance of marked chromosomes in *Drosophila*. He studied aneuploidy of the sex chromosomes and, in a classical paper, used the hypothesis of nondisjunction to obtain proof of the chromosomal basis of inheritance. He observed rare exceptions to the regular segregation of sex-linked mutations. Exceptions were of two kinds: matroclinous (which inherited two X chromosomes from the mother and a Y from the father), and patroclinous (which received no maternal sex chromosome but an X from the father; see Fig. 3.3). The resulting XXY and XO types were female and male respectively. He attributed these exceptions to a failure of separation of the paired chromosomes at meiosis in the mother. Bridges also summarized cases of exceptions to the normal sex-linked pattern of inheritance reported in other organisms which were also consistent with the hypothesis of nondisjunction. The publication of this paper led to a rapid acceptance of nondisjunction as an aberrant meiotic process leading to aneuploidy. It should be noted, however, that the genetic markers which enabled the aneuploidy to be detected did not permit a definite conclusion to be drawn as to the nature of the meiotic error which produced it.

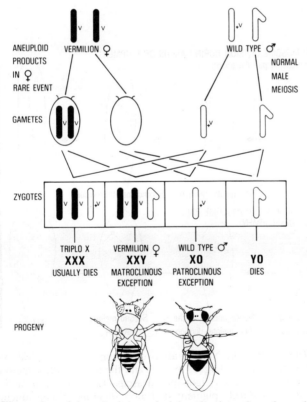

Fig. 3.3. Illustration of the principle underlying the detection of sex-chromosome aneuploids in *Drosophila* using genetic markers.

Markers located at the centromere allow a distinction to be made between errors occurring at the first and second meiotic divisions, but the various possible first division errors at meiosis could not be distinguished.

Aneuploid detection in Ascomycete fungi

In ascomycetes such as *Neurospora* and *Sordaria*, the meiotic products are retained within a specialized cell called an ascus. Each ascus is derived from a single diploid nucleus known as an ascus initial and contains eight ascospores. The ascospores are produced by a classical meiosis followed by the mitotic division of each meiotic product.

Either nondisjunction at, or nonconjuction prior to, the first meiotic division, will lead to the simultaneous production of an equal number of disomic ($n + 1$) and nullisomic ($n - 1$) products (Fig. 3.4). Mitchell *et al.* (1952) were the first to use suitably marked strains of *Neurospora crassa* to detect disomic ascospores in asci which also contained four, presumably

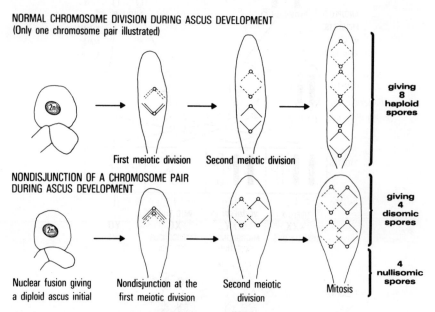

NORMAL CHROMOSOME DIVISION DURING ASCUS DEVELOPMENT
(Only one chromosome pair illustrated)

First meiotic division   Second meiotic division   giving 8 haploid spores

NONDISJUNCTION OF A CHROMOSOME PAIR DURING ASCUS DEVELOPMENT

Nuclear fusion giving a diploid ascus initial   Nondisjunction at the first meiotic division   Second meiotic division   Mitosis   giving 4 disomic spores   4 nullisomic spores

Fig. 3.4. *Upper diagram*: Normal ascus development in a typical Ascomycete. The diploid ascus initial undergoes meiosis and the four resulting meiotic products divide mitotically to give an eight-spored ascus. In *Neurospora* or *Sordaria* the spores are normally darkly pigmented. *Lower diagram*: Nondisjunction for one chromosome pair during the first meiotic division. The resulting ascus contains four disomic spores and four inviable nullisomic spores.

nullisomic, abortive spores. Plate 3.1 illustrates this type of ascus from a cross of two buff mutants of *Sordaria brevicollis*. The mutants are closely linked alleles of a spore colour locus on linkage group II. They complement each other, so that any ascospore which inherits both alleles through an abnormal event can be detected as a non-mutant black spore.

The disomic nature of the spores, both in *Neurospora* and *Sordaria*, can be readily demonstrated by dissecting them out of the ascus and allowing them to germinate. The disomic nucleus is unstable and one or other of the extra chromosomes is soon lost. The fungal colony thus established from a disomic spore rapidly becomes a heterokaryon containing two genetically different types of nuclei. Despite the slight complication that there is an opportunity for mitotic recombination to occur in the $n + 1$ nucleus before it breaks down (Pittenger and Coyle 1963; Coyle and Pittenger 1965), it is a comparatively simple task to demonstrate the presence of the complementary linkage groups which permitted the detection of aneuploidy in the first place.

Asci such as these provide compelling evidence for an error at the first division of meiosis. However, both the number and distribution of spores in this type of ascus is equally consistent with the underlying defect being nondisjunction or nonconjunction. The relative proportions of the asci arising from nonconjunction, in which synapsis is either absent or not maintained for the normal period, and nondisjunction, in which paired chromosomes fail to separate, has not been determined in Ascomycetes.

Several methods have been published for detecting aneuploidy in fungi, which rely on the plating of random meiotic products under conditions in which only aneuploids or prototrophic recombinants can grow (Griffiths and Delange 1977; Parry 1977; Parry *et al.* 1979a; Sora *et al.* 1982). Mutations conferring growth requirements are arranged along the chromosome in such a way that multiple crossing over, necessary to generate the prototrophic recombinants, will be extremely rare, and so the majority of spores growing on selective medium will be aneuploid. These methods are potentially a very valuable way of studying aneuploid induction, but they select preferentially for aneuploids which are formed as a result of nonconjunction. The extent of this bias varies and depends on the length of the chromosome segment in which the absence of recombination is being selected.

DISTINGUISHING NONCONJUNCTION FROM OTHER CAUSES OF MEIOTIC ANEUPLOIDY IN *DROSOPHILA*

It is in *Drosophila* with its genetically well marked chromosomes that experimental evidence distinguishing nonconjunction from other causes of aneuploidy has been obtained. The underlying rationale for this distinction is that nonconjunction arises from asynapsis or desynapsis and asynaptic chromosomes are necessarily non-recombinant, whilst desynaptic chromosomes are very likely to show reduced recombination or none at all. The extent therefore to which aneuploids are also non-recombinant can be used as an indication of the relative frequency of nonconjunction.

In 1936, Weinstein published an algebraic method for calculating the frequency of meiotic division with a given number of exchanges in a particular bivalent. The method assumes that the linkage group is marked throughout its length and that there is no chromatid interference between exchanges. Provided that the marked intervals are small, so that there is a negligible frequency of undetected doubles within an interval, it is possible to calculate the frequency of bivalents with no exchanges. From estimates made by many workers, it would appear that about 5 per cent of meiotic divisions in females of *Drosophila melanogaster* have no exchanges along the X chromosome. Since sex-chromosome aneuploids are normally formed with a frequency much lower than this, it follows that the zero-exchange bivalents are normally able to undergo regular disjunction.

Merriam and Frost (1964) investigated the crossover frequency in a multiply marked X chromosome in an attempt to separate nonconjunction from other possible modes of origin. They applied Weinstein's method to calculate the relative frequency of exchanges and found that in the meiotic divisions giving rise to matroclinous exceptions 0, 1, 2, and 3 exchanges occurred with a frequency of 26, 24, 47.5, and 2 per cent respectively. This compared with frequencies in controls of 4.6, 65.7, 28.7, and 1 per cent. The striking features of these results are the increase in zero-exchange tetrads, the dearth of single exchanges, and the higher proportion of multiply exchanged bivalents amongst exceptions. Merriam and Frost suggested that both modes could be operating, some aneuploids arising by failure of separation in multiple-exchange bivalents, as suggested by Bridges, others arising by nonconjunction of chromosomes, as suggested by Sturtevant and Beadle (1936).

At this point it should be noted that in *Drosophila* the frequency of aneuploidy in progeny of individuals who are themselves aneuploid is greatly increased. Bridges called the exceptions arising in this way 'secondary' exceptions and the process giving rise to them he called 'secondary nondisjunction'. Secondary nondisjunction is due almost entirely to an increase in the frequency with which non-exchange chromosomes are found in aneuploid meiotic products. Recently however, Luning (1982*a,b*) has examined the frequency of recombination during the process of secondary nondisjunction and shown that recombination in the distal region of the X chromosomes, marked by yellow and white, is normal. The experimental analysis of secondary nondisjunction will form the basis of a later section of this chapter.

DETECTING ANEUPLOIDY ARISING FROM PRECOCIOUS CENTROMERE DIVISION

Aneuploids that arise either from nonconjunction or nondisjunction will be formed at meiotic divisions in which half of the products are disomic ($n + 1$) and the other half nullisomic ($n - 1$). Thus the type of ascus illustrated in Plate 3.1 most probably originated in one of these ways. It is not, however, the only meiotic route to aneuploidy as can be seen in *Sordaria* by the detection of asci in which only half the meiotic products are aneuploid. Such an ascus can be seen in Plate 3.2. In asci such as these the disomic and nullisomic products have segregated away from each other at the first meiotic division and are located together with normal haploid spores, in opposite halves of the ascus. Threlkeld and Stoltz (1970) were the first to describe asci of this type in *Neurospora crassa*. They were detected in crosses of two complementary pantothenic acid requiring strains. The presence of either mutation produced an effect on ascospore colour, thus changing the normal black spore to a pale type. Exceptional ascospores which inherited both alleles were identified by the restoration of full colour. The authors suggested that asci similar in type to that shown in Plate 3.2 arose through premature or precocious centromere division of one homologue in a bivalent.

In any organism in which tetrad analysis is impossible the distinction between premature centromere division and nondisjunction at the first division cannot be made on genetic grounds. This is because both events give rise to aneuploids which are heterozygous for any centromere markers. Only the proportion of aneuploid meiotic products (50 per cent in the case of premature division; 100 per cent for first-division nondisjunction) distinguishes the two phenomena. Thus, Merriam and Frost (1964) pointed out that the experimental evidence for nondisjunction in *Drosophila* was equally compatible with a meiotic division in which the first division was equational for one half-bivalent (indicating premature centromere division) rather than reductional. None the less, evidence that precocious centromere division does occur has been obtained in organisms other than fungi, using cytogenetic methods.

Hughes-Schrader (1948) studied the behaviour of the single X chromosome in spermatocytes of the mantid *Humbertiella indica*. She observed that the X chromosome lagged behind the other chromosomes at anaphase and was expelled from the spindle. This was a consequence of its being an univalent. In some meiotic divisions, however, the X chromosome was no longer an univalent but behaved like a bivalent through the premature division of its centromere. Expulsion did not then occur.

Polani and Jagiello (1976) in their study on the maternal age effect in mice recorded the frequency of cells at metaphase II with single unpaired chromatids. Out of 1142 cells examined from the strain C57 Bl, 17 had a single chromatid and a hypohaploid chromosome number while one had a count of 20 + ½. In the strain CFLP the corresponding figures were six and one out of a total of 391 cells. There were no cases where a metaphase II cell had two unpaired chromatids. Polani and Jagiello called the phenomenon giving rise to the single chromatids 'predivision' whilst Hansmann and El-Nahass (1979) used the term 'presegregation' for a similar observation. Out of a total of 3338 oocytes, Hansmann and El-Nahass observed nine with 19 + ½ and six with 20 + ½ chromosomes.

In particular circumstances the division of the centromere at meiosis I might be quite frequent. Michel and Burnam (1969) studied the behaviour of univalents in double-trisomic maize and observed the equational division of an univalent in 13.1 per cent of all cells. Guillemin (1980) has also reported precocious centromere division in a trisomic newt. Equational division may therefore have a high frequency following nonconjunction.

There is evidence, however, that prematurely divided unpaired chromatids do not necessarily yield disomic products. Ostergren (1947) reported that, in the grass *Anthoxanthum*, univalent chromosomes sometimes divided at the first division but when this occurred, the resulting chromatids were unable to divide at anaphase II and were lost from the spindle.

DETECTING ANEUPLOIDY ARISING AT THE SECOND MEIOTIC DIVISION

Among the exceptions reported by Bridges (1916) in *Drosophila* there were 18 cases where a matroclinous exception was homozygous for one of two linked genes whilst the mother had been heterozygous for both. Bridges attributed these to the occurrence of crossing-over at the four-strand stage followed by failure of the chromatids to separate at the second, equational, division of meiosis. Thus these matroclinous exceptions received one crossover and one non-crossover chromatid from their mother. They were called 'equational' because of the supposed failure of the equational division. At this time the location of the centromere was not known but when this information became available, Anderson (1929) observed that the frequency with which markers were found to be homozygous increased as the distance of the markers from the centromere increased. Genetic markers located at the centromere were nearly always heterozygous and this was not to be expected if the equational exceptions arose through nondisjunction at the second division. Anderson (1929) therefore concluded that the absence of homozygosity at the centromere was an indication that the exceptions arose from nondisjunction at the first division, the homozygosity of the markers being explained, as Bridges (1916) had postulated, by crossing-over. He obtained further support for this idea by detecting exceptionals which had inherited the reciprocal products of crossing over. These would not be expected if nondisjunction had occurred at the second division. Furthermore Anderson (1929) showed that if the equational exceptions arose exclusively at the first division, then, provided the second division was normal, they would be expected to occur with twice the frequency of exceptions inheriting reciprocally recombinant chromatids. The basis of this expectation is presented diagrammatically in Fig. 3.5. Anderson (1929) observed that 7.03 and 3.55 per cent of 1144 exceptions were equational and reciprocal recombinants, respectively, and this implied that very few equational exceptions could have arisen at the second division. He derived this conclusion from work on a special stock of *Drosophila* which showed a high spontaneous frequency of aneuploidy. This stock was later shown (Lewis 1951) to contain a chromosome rearrangement. Merriam and Frost (1964), however, arrived at a similar conclusion after studying a strain of *Drosophila* showing normal spontaneous levels of aneuploidy. They estimated that the second division nondisjunction frequency in females was only 1/75 000. It thus appears, in *Drosophila* females, that the majority of exceptions, whether regular or equational, arise from errors at the first meiotic division. In *Drosophila* males, however, the situation may be different. From a cross of XXY females to males whose X chromosome was marked at the white eye locus, Lamy (1949) reported 32 exceptional offspring each with two paternal X chromosomes, in a total of 143 313 progeny. One explanation is that the exceptions arose through nondisjunction at the second division in the male parent. If this is the explanation, it would indicate that the frequency of

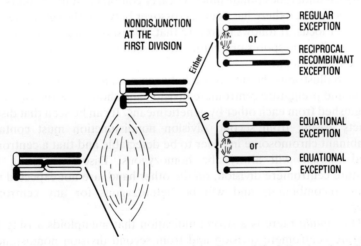

Fig. 3.5. Disomic gamete types expected in *Drosophila* following first division nondisjunction of a bivalent with a single crossover. Assuming that the segregation of the chromosomes into the oocyte is random, and the centromeres divide normally at the second division, then equational exceptions will be recovered twice as often as reciprocally recombinant exceptions.

this event in males is much higher than in females. This observation has never been followed up and other possible explanations, such as mitotic nondisjunction in the developing zygote would have to be excluded before this conclusion could be drawn with certainty.

The rarity of second division errors characteristic of the *Drosophila* female certainly does not hold for other organisms. In both *Neurospora* and *Sordaria*, ascus analysis yields strong evidence for the occurrence of second division nondisjunction. Threlkeld and Stoltz (1970) described asci which they believed had arisen in this way in *Neurospora* and Plate 3.3 shows an ascus from the *Sordaria* system which can be explained by nondisjunction at the second division. Asci arising from this cause are superficially very similar to those arising from premature centromere division; the two types are distinguished by the sequence of the spores and by the behaviour of centromere-linked markers (if any). In *Neurospora*, an ascus arising from a second-division error will contain two disomic and two nullisomic spores in the same half of the ascus. In *Sordaria* the positions of the two centrally located spore pairs are often switched during development of the ascus. This phenomenon, known as 'spindle overlap', means that spore sequence in *Sordaria* cannot always be used to infer the origin of the aneuploid spores. With respect to the spore sequence, the asci photographed in Plates 3.2 and 3.3 are clearly different and this conclusion is not affected by the occurrence of spindle overlap. They differ too in another respect. Nondisjunction at the second division is detected in experimental systems such as these, only when the two chromatids of the dyad which undergoes nondisjunction carry complementary markers. This comes about when there has been a crossover between the centromere and the marker locus. It follows therefore that at least one of the chromosomes present in disomic spores originating from second-division nondisjunction must be recombinant.

Fig. 3.6 illustrates the difference between nondisjunction at the second division and premature centromere division and shows how the two can be distinguished from each other by genetic means. It can be seen that disomic products arising from second division nondisjunction must contain a recombinant chromosome in order to be detectable and that a centromere located marker will always be homozygous. Disomics arising from premature centromere division, on the other hand, are not expected to be obligate recombinants and will be heterozygous for any centromere marker.

In *Drosophila* there is a strong indication that aneuploids arising from premature centromere division and from second division nondisjunction can be the result of very similar cellular defects. This conclusion arises from work with two particular meiotic mutants. Mason (1976) described a semidominant autosomal mutant (*ord*) which greatly increases the frequency of regular and equational exceptions. The mutation affects

Premature centromere division

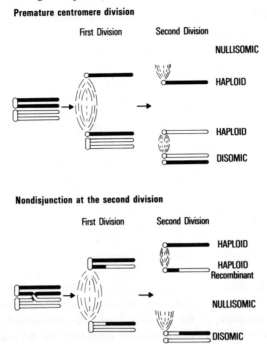

Fig. 3.6. Comparison of meioses in which aneuploids are generated by (a) premature centromere division and (b) nondisjunction at the second meiotic division. In (a) the disomic products will be heterozygous for any centromere-located genetic markers, whilst in (b) the disomics will be homozygous at the centromere. If detection of the disomic is dependent on complementation of distal loci then disomics originating from (b) will be obligate recombinants.

aneuploid frequencies in both males and females. Cytological examination (Mason 1976; Goldstein 1980) revealed that precocious separation of centromeres occurred with a high frequency in homozygotes. Although the mutation also affected recombination frequency in females, Mason ascribed this to a secondary consequence of the defect and proposed that *ord* was defective in an early acting meiotic function which served to hold the chromatids together. Davis (1971) had earlier described a very interesting mutant, *mei - S332*, which had marked consequences on the second meiotic division. The mutation, again semidominant and autosomal, resulted in the production of about 20 per cent nullisomic and 12 per cent disomic gametes. Davis showed that in suitably marked heterozygous females, all the matroclinous exceptions in a sample of 179, and 530 exceptions in another sample of 555, were homozygous for a centromere marker. This constitutes clear genetic evidence that the mutation affects chromosome behaviour at the second meiotic division, and contrasts with the finding that the frequency of spontaneous nondisjunction at the second division in normal flies is very low. Again, just as in the case of *ord*, there was clear cytological evidence for premature separation of the centromere in many cells. Baker *et al.* (1978) suggested that the function of the non-mutant *mei - S332* gene was to delay the separation of centromeres during all cell divisions. Goldstein (1980) has suggested that both *ord* and *mei - S332* are defective in sister chromatid cohesion and the difference between the mutants lies in the time at which the genes function during meiosis. *Mei - S332*, on this view would be a later-acting gene than *ord*.

It appears therefore that second-division nondisjunction can arise from a defect resulting in premature separation of centromeres provided the separation is delayed until after anaphase I. When this happens the disassociated centromeres assort independently at the second meiotic division, thereby generating aneuploidy. Maguire (1982) has recently shown, in maize, that pachytene synapsis of the centromere region is important in providing for sister centromere association until anaphase II.

It is possible to envisage another way in which second-division aneuploids may arise. A centromere defect resulting in delayed separation of chromatids at the second meiotic division would have precisely the same genetic consequences, so far as aneuploidy was concerned, as the defect present in *mei - S332*. It would appear however that no mutant with this characteristic has been described.

DETECTING ANEUPLOIDY ARISING FROM CHROMOSOME REPLICATION ERRORS

In a sense, there are just two ways in which the extra chromosome, present in a hyperploid gamete or individual, can arise. It may be formed either through a segregational error at a cell division or through a chromosome replication error.

There is no doubt that in mitotically dividing cells, extra replication of the whole chromosome complement can occur which results in polyploidy. White (1935) was the first to describe, in *Locusta migratoria*, the induction, with X-rays, of chromosomes having four chromatids attached to a single centromere. White coined the term 'diplochromosomes' for these structures. Levan (1939a) also induced diplochromosomes in plant material using plant hormones. As the doubling of the chromosomes occurred during interphase, it could be distinguished from colchicine-induced doubling associated with cell division failure. Levan showed that after mitotic division of diplochromosomes the chromosome number of the daughter cells was doubled. In many animals endoreduplication occurs in some of their somatic cells, as a normal feature of development, to generate polyploidy.

Barber (1940) reported the suppression of meiotic cell division of

Fritillaria meleagris, by use of elevated temperature. The diplochromosomes which were formed, associated as diplobivalents. Barber suggested that these arose from repeated chromatid division without nuclear division. When the temperature inhibition was lifted, the centromeres of each diplochromosome underwent two successive divisions which resulted in the formation of daughter cells with a polyploid number of normal-looking chromosomes. Again in this case, the entire chromosome set underwent reduplication to generate a polyploid, as opposed to an aneuploid, meiotic product. In some organisms chromosome reduplication occurs spontaneously as a normal part of the life cycle. In parthogenetic organisms such as the stick insect *Carsinsius morossus*, Koch *et al.* (1972) reported that the whole genome underwent extra replication prior to meiosis.

In all these cases the whole chromosome complement in affected cells was duplicated. Evidence that single chromosomes can undergo extra replication generating aneuploidy is much more meagre. The first reported case did not stand up to reanalysis. Sturtevant and Dobshansky (1936) investigated a mutant called 'sex ratio' in *Drosophila pseudobscura*. In this strain, affected males yielded almost entirely female progeny. Sturtevant and Dobzhansky reported that in spermatocytes of 'sex ratio' males, the X chromosome was split longitudinally so that it consisted of four chromatids. The resulting meiotic products therefore all contained X chromosomes, which accounted for the distorted sex ratio. Novitski *et al.* (1965) and Policansky and Ellison (1970), however, both re-investigated the cytology of this mutant, and were unable to find any evidence of extra replication of the X chromosome.

There have been some reports of the spontaneous doubling of univalent chromosomes which subsequently segregate normally on the division spindle (Battaglia 1964; Ford 1971). This phenomenon is easily confused with premature centromere division, the difference being that in premature centromere division the segregating elements are single chromatids which may fail to divide at anaphase II and as a consequence be lost from the spindle (Ostergren 1947), whilst in the case of extra replication, the segregating element is a dyad which would be expected to behave normally at the second meiotic division.

Cytological evidence for extra replication leading to aneuploidy is very poor; the two strongest lines of evidence for it come from genetic analysis. Cell division errors, which generate hyperploid daughter cells, will, at the same time, give rise to an equal number of hypoploids. Extra replication, on the other hand, will give rise exclusively to hyperploidy. Once again it is organisms in which all the products of a single meiosis can be examined which provide the clearest evidence for this type of meiotic error.

Plate 3.4(a) shows an ascus from the *Sordaria brevicollis* system in which the four disomic black spores are produced together with four buff spores. By the use of ascus dissection and by standard analysis of the disomic spores it can be shown that asci such as these contain 'extra' replicas of the marked linkage group. At the end of the mitosis following meiosis there are twelve copies of the linkage group distributed between eight nuclei. Other asci in which there are extra chromosomes are also found. Plate 3.4(b) shows an ascus with two black and six buff spores in which the black spores are again disomic. Bond (1976) suggested that these asci contain two buff disomic spores in addition to the black disomics. The buff disomics contain two copies of linkage group II per nucleus, which are not marked with the complementary buff mutations because crossing-over has resulted in the second division segregation of the buff locus. Recent experiments (Bond, unpublished) have confirmed that this type of ascus does have a crossover in the centromere–buff interval as predicted. It seems clear therefore that both in asci with two black and four black spores, the asci originated from an ascus initial which was not diploid as in the normal case, but was trisomic.

Both of the ascus types illustrated in Plate 3.4 could have originated either through extra replication of the marked chromosome or through mitotic nondisjunction immediately prior to meiosis. This latter possibility is discussed further on p. 47. Mitotic nondisjunction, however, is an inadequate explanation for another type of ascus, first described by Case and Giles (1964). They dissected 1457 asci, each with eight fully pigmented and viable ascospores, from a *Neurospora* cross segregating for several markers on linkage group VI. By this procedure they eliminated from consideration asci in which a meiotic division error had occurred, since these events generate some unpigmented, inviable spores. None the less they detected nine asci with disomic spores. Each ascus contained one disomic and seven haploid spores. In three of the cases the disomic ascospores contained chromosomes which had undergone crossing-over at meiosis. Nondisjunction at the first mitotic division after meiosis would have generated one disomic and one nullisomic abortive, spore. Since the nine asci contained no abortive spores the possibility must be considered that the 'extra' chromosome arose through a replication error. No other simple explanation comes to mind to account for asci such as these.

The other piece of evidence for extra replication comes from work on a meiotic mutant of yeast. Moustacchi *et al.* (1967) reported a dominant mutation called 'abnormal meiosis' which was characterized by a high frequency of aberrant segregation. Instead of the expected Mendelian 2+ : 2 mutant segregation they observed high frequencies of 3+ : 1 mutant and 4+ : 0 mutant ratios in crosses segregating for a variety of genetic markers. Subsequent analysis eliminated several explanations such as back mutation, polyploidy and supernumerary mitoses, and the authors concluded that extra replication of chromosomes, which resulted in a high frequency of aneuploidy, was the underlying mechanism.

On the basis of this slim evidence it would probably not be worth considering extra replication as a significant factor in generating aneuploidy, but for the fact that it is a mechanism which generates only hyperploid products. As we have seen in Chapter 2, trisomic zygotes in man and other mammals are detected more frequently than the corresponding monosomics. This inequality is likely to be due to preimplantation death of the monosomic zygote rather than to impaired functioning of the nullisomic gamete, but there remains a further possibility that there may be a real excess of trisomic embryos. Furthermore, in man, there is a well authenticated increase in the frequency of trisomics with increase in maternal age (see Chapter 4) and although many hypotheses have been advanced to explain the age effect, no one explanation has gained universal acceptance. Maudlin and Fraser (1978*b*) have studied the frequency of aneuploidy in one-cell mouse embryos. The technique permitted maternal and paternal chromosomes to be distinguished from each other and they reported a significant increase in aneuploidy of the maternal set in older mice. The increase in aneuploidy, however, occurred only for the trisomics, no corresponding age effect on the frequency of monosomic embryos being detected. Martin *et al.* (1976) had previously reported an increase in hyperploid MII oocytes in mice of an age corresponding to those used by Maudlin and Fraser. In both experiments the number of hyperploid cases reported was small, but should these findings be confirmed with more extensive data, it will obviously be necessary to consider meiotic errors which generate hyperploidy only. It might then be worth considering whether the age effect in females can be explained by postulating an increased probability of an extra replication of chromosomes as the dictyate stage continues in the oocyte.

DETECTING ANEUPLOIDY ARISING BY CHROMOSOME LOSS

Several factors contribute to a comparative dearth of information about events specifically associated with chromosome loss. The comparatively high frequency with which chromosomes are lost artefactually from a cell during preparation for cytological examination, severely limits the contribution of cytogenetics, whilst the inviability of the meiotic products in lower organisms and the monosomic zygotes in higher organisms, imposes a similar limitation on genetic studies.

In *Drosophila* the loss of a sex chromosome at meiosis can result in the formation of a viable XO zygote. The processes giving rise to chromosome loss can therefore be studied in this case. In the earliest experiments on sex chromosome aneuploidy in *Drosophila* (Bridges 1916; Safir 1920) it was observed that chromosome losses (giving patroclinous exceptions) occurred with a greater frequency than chromosome gains (giving matroclinous exceptions). This was unexpected if losses originated only through nondisjunction and indicated either that the disjunctive products were not recovered equally or that errors might exist giving rise specifically to nullisomic gametes.

The observation of monosomic progeny is several steps removed from the meiosis giving rise to the gametes involved and there is a real possibility therefore that the excess of sex chromosome loss could arise from any one of several post-meiotic events, such as gamete dysfunction or elimination of chromosomes from the zygote after fertilization.

Sperm dysfunction could generate an excess of chromosome loss types if the development of the disomic meiotic product were impaired in some way whilst that of the complementary nullisomic product were not. There is evidence that in certain circumstances this may be the case. Lindsley and Sandler (1958) studied aneuploidy in a *Drosophila* stock carrying an attached X-Y chromosome and a deleted X. They recovered the deleted X in more than half of the progeny. There was no appreciable zygote mortality and, since meiotic loss alone can never result in the recovery of a homologue in more than 50 per cent of gametes, they concluded that impaired development of the X-Y bearing sperm was responsible in this case. Sperm dysfunction was proposed by Hartl *et al.* (1967) and by Nicoletti *et al.* (1967) to account for segregation distortion in *Drosophila*.

In a direct development of the work of Lindsley and Sandler (1958) on *Drosophila*, Miklos (1974) proposed that successful sperm development is dependent, in many species, on pairing of chromosomes. If chromosomes are unpaired in meiosis then development of a gamete inheriting any unpaired chromosome is curtailed due to the presence of 'unsaturated pairing sites' on the chromosome. The hypothesis could provide a simple explanation for the preferential recovery of nullisomic gametes. Such gametes would not contain chromosomes with unsaturated pairing sites and their development should not therefore be affected. Disomic gametes, on the other hand, could be impaired in development if the disomic condition arose from an unpaired state. On this hypothesis, the disomic and nullisomic products of nonconjunction would not be expected to be recovered with equal frequency. Miklos *et al.* (1972) and Peacock *et al.* (1975) gave a direct demonstration of impaired development when they showed that as the reciprocal aneuploid products were recovered with increasing inequality, there was a corresponding increase in spermatid breakdown as visualized by the electron microscope.

Genetic experiments cannot always discriminate between sperm dysfunction and chromosome loss at meiosis. Thus Sandler and Braver (1954) carried out experiments in which they systematically varied the amount of heterochromatin on the X chromosome by using deletions of varying size. They showed that as the size of the deleted segment increased, X-loss also increased so that O gametes were recovered in frequency far in excess of the complementary XY type. In addition Y-bearing sperm were recovered less often than deleted-X-bearing sperm and this inequality too increased with the size of the deleted segment. Sandler and Braver (1954) suggested that chromosome loss increased with reduced pairing. This explanation was, however, questioned by Cooper (1964) and Peacock (1965), who were unable to demonstrate cytologically any chromosome loss at meiosis, but were able to show that XY and O cells occurred with a frequency similar to that with which the X and Y were unpaired. Peacock *et al.* (1975) showed that sperm dysfunction subsequently occurred, development of the XY-bearing sperm being impaired. The 'saturated pairing' hypothesis accounts for the preferential recovery of O sperm (they contain no unpaired sites and develop normally) and for the deficiency of Y-bearing sperm compared to deleted-X-bearing sperm (the latter carry fewer unpaired sites because of the deleted segment).

It has been suggested that monosomic zygotes can also arise through post-fertilization loss; Russell (1961) and later Russell and Montgomery (1974) demonstrated that XO mice occurred spontaneously with a much greater frequency than XXYs. Of the two alternative explanations for this disparity—(i) meiotic loss leading specifically to nullisomic gametes or (ii) post-fertilization errors giving XO zygotes, Russell (1961) favoured the latter, although a meiotic origin for some was not completely ruled out. The same workers also showed that loss of the paternally derived sex chromosome, yielding X^mO mice, occurred with a much higher frequency than loss of the maternally derived X, yielding X^POs.

These factors complicate the genetic study of chromosome loss at meiosis in higher organisms. Cytogenetic studies, however, indicate that loss of a chromosome can occur during cell division, particularly if the chromosome is an univalent. Hughes-Schrader (1948) for example studied the expulsion of the univalent X chromosome from the spindle of *Humbertiella indira*. It was suggested that expulsion was a consequence of a spindle connection to only one pole and cytoplasmic streaming across the equator of the spindle which tended to carry the unanchored chromosome out of the spindle. Ford (1971) studied the irregular behaviour of univalents in *Leucopogon juniperinus* and observed that they were attached to separate reduced 'mini-spindles'.

Errors of chromosome distribution leading to aneuploidy have also been observed following the formation of multipolar spindles (Fankhauser 1934; Böök 1945; Barthelmess 1957; Nur 1963). Böök (1945) subjected fertilized eggs of the salamander, *Triton*, to cold-shock treatment before the first cleavage division, and this induced aneuploidy in a variety of ways. The aneuploids arising from multipolar spindles were mosaics which had simultaneously lost several chromosomes. Böök called these 'multiform' aneuploids. There are, however, other reports where multipolar spindles have not generated aneuploid daughter cells (Teplitz *et al.* 1968; Pera and Schwarzacher 1969).

In *Drosophila*, attached $\widehat{XY}$ stocks ($\widehat{XY}$/O), $\widehat{XY}$, and O sperm are produced with some excess of O sperm. There is no sperm dysfunction in this case, which is consistent with the saturated pairing hypothesis because it has been shown that the $\widehat{X}Y$ folds back to pair on itself (Cooper 1964), so that no gametes contain chromosomes which have been unpaired. Hardy (1975) examined sperm heads in $\widehat{X}Y$/O stocks with the electron microscope, and was able to observe directly the elimination of the $\widehat{X}Y$ chromosome into a micronucleus which was thought to be lost from the sperm later in development. In this stock therefore the deficiency of $\widehat{X}Y$ offspring was traced, more or less directly, to chromosome loss at meiosis.

Chromosome loss has also been detected cytologically in Chinese hamster oocytes. Sugawara and Mikamo (1980) examined MII oocytes which had been treated with colchicine and found a significant increase in aneuploids, some of which originated through the lagging of chromosomes at anaphase I. Such chromosomes were not included in either the oocyte or the polar body and could be seen as degenerating chromatin masses in the region between the oocyte and polar body.

Chromosome loss has also been studied using mutants in various organisms. In some cases mutants which affect loss quite specifically have been isolated; no concomitant increase in chromosome gain can be detected. For example, Gelbart (1974) has described 'mitotic loss inducer' in *Drosophila*, a recessive mutation which increased mitotic, but not meiotic, loss of the X chromosome in females. The process giving rise to haplo-X tissue did not produce a corresponding trisomic-X sector.

From several of the studies using meiotic mutants it has emerged that the centromere region of the chromosome might be important in influencing the frequency of chromosome loss. Spieler (1963) described a strain of *Drosophila* which gave a high frequency of patroclinous exceptions. Genetic analysis revealed that the high frequency of chromosome loss was the result of a factor at the centromere end of the chromosome. Baker (1975) investigated a recessive mutation in *Drosophila* called paternal loss (*pal*). The mutation affected only males and paternally derived chromosomes in their progeny. An increased frequency of loss for all such chromosomes in the genome was observed, and Baker concluded that *pal* males produced defective chromosomes which had an increased probability of being lost during the early cleavage division. The frequency with which *pal* males showed sex-chromosome loss in their progeny was influenced by their maternal genotype, an observation which led Baker to conclude that some, if not all, of these apparently meiotic exceptions originated from loss of the paternal chromosomes in the zygote. Not all chromosomes were equally sensitive; it was shown, for example, that some X chromosomes were more sensitive than others to the effect of *pal*. This differential sensitivity resided at the centromere region of the X chromosome although *pal* itself is a second chromosome mutation. Baker favoured the hypothesis that the *pal*+ product interacted with the centromere region of chromosomes, so that, in the mutant, which lacked the product, loss came about through a centromere defect which affected chromosome movement.

In yeast, too, there is evidence that specific chromosome loss at mitosis can be brought about by gene mutation. Haber (1974) and Liras *et al.* (1978) investigated a recessive gene (*chl*) causing chromosome loss. One per cent of mitotic cell divisions resulted in loss of chromosome III. There was a smaller effect on linkage groups VIII and XVI whilst six other linkage groups tested showed no increased loss. There was an indication that the mutation also increased the frequency of meiotic loss because a large proportion of tetrads contained inviable spores. From the failure to demonstrate the simultaneous production of disomic spores in such tetrads it can be inferred that chromosome loss was not brought about through meiotic nondisjunction. Liras *et al.* (1978) also reported that mitotic chromosome loss sometimes occurred in conjunction with mitotic crossing-over, a finding that had also been reported earlier by Campbell *et al.* (1975) in studies on trisomic yeast. Campbell and Fogel (1977) further investigated the relationship between mitotic loss and mitotic recombination in experiments in which they selected for mitotic recombination in an interval near the centromere and, separately, in a distal interval. They demonstrated that selection for mitotic recombination near the centromere also selected for increased loss, whereas this relationship disappeared for distally selected recombination. They concluded that this finding was consistent with a causal relationship between chromosome loss and centromere-located mitotic recombination, and so, once again, there is evidence for the involvement of the centromere region in generating chromosome loss. It should be noted that in the system studied by Campbell and Fogel (1977), chromosome loss arising from a nondisjunctional event was not excluded. The study of meiotic mutants indicates that some defective cellular processes can bring about chromosome loss, and the centromere regions of chromosomes, in particular, are implicated in these processes.

DETECTING ANEUPLOIDY ARISING BEFORE MEIOSIS

The possibility that some aneuploids arose mitotically prior to meiosis was carefully considered by the early *Drosophila* workers, e.g. Safir (1920). If a substantial proportion of *Drosophila* exceptions were mitotically derived they would be distributed non-randomly between individual flies. This is because a mitotic non-disjunction event, unlike a meiotic one, will give rise to a cluster of aneuploids through further mitotic division. The progeny of *Drosophila* can be analysed for the presence of clusters by analysing the progeny of individual females to determine whether or not the aneuploids are randomly distributed. The result of this sort of analysis in *Drosophila* gave no evidence however for a clonal distribution, and the conclusion was reached that the exceptions originated during meiosis.

A similar non-random distribution of mitotically derived aneuploids is expected in *Sordaria* and *Neurospora*. In both these organisms there are thought to be no mitotic divisions of the diploid nucleus. The possibility of a mitotic error therefore refers to misdivision of the haploid nucleus, an event which will generate disomic and nullisomic products. The initial error will, by further mitoses, be expected to give rise to clusters of disomic and nullisomic nuclei. When these fuse with haploid nuclei immediately prior to meiosis, trisomic and monosomic ascus initials will be formed. Of these only the trisomic nucleus gives detectable aneuploidy. The other leads to nullisomy and spore abortion. The abortive spores can be seen but since they are inviable no genetic analysis can be carried out, and they are confounded with abortive spores arising from other developmental aberrations.

The trisomic ascus initial, on the other hand, will give rise to an ascus with disomic and haploid products like the one photographed in Plate 3.4. Since asci are formed with discrete fruiting bodies, each containing up to 300 asci, clusters of aneuploids derived from an early mitotic event should be detected very easily. However, no clustering on asci containing four disomic : four haploid products has ever been reported. In both *Neurospora* and *Sordaria* the absence of clusters can be explained through the known tendency for a disomic nucleus to lose the extra chromosome (see p. 33). The detection of mitotic recombination in cultures established from germinated disomic spores (Pittenger and Coyle 1962; Threlkeld and Stephens 1966; Bond, unpublished) indicates that the aneuploid nuclei are not completely unstable (they exist long enough for a recombination event to occur) but the absence of clustering could be explained by the hypothesis that disomic nuclei are not stable enough to allow clusters to form. According to this idea the majority, perhaps all, of the mitotically originating nuclei which eventually give rise to aneuploid meiotic products are those which are 'rescued' by fusion with a haploid nucleus immediately prior to meiosis to form a trisomic ascus initial.

In mammals, the clustering of mitotically derived aneuploidy would be difficult to detect in genetic tests due to small litter sizes. Some data on nondisjunction in spermatogonial divisions are available, however, from experiments in which spermatogonia have been treated with X-rays with a view to increasing meiotic aneuploidy. Russell and Montgomery (1974) irradiated spermatogonia with both single and fractionated doses of X-rays, and recorded the subsequent frequency of XO females among the F_1 offspring. They were unable to detect any increase and concluded that most XO females arose, not from meiotic errors, but after sperm entry into the egg through the loss of the paternal sex chromosome. Ford *et al.* (1969)

also found no evidence for X-ray induced aneuploidy when 1200 R in two equal fractionated doses was given to spermatogonia. Speed and Chandley (1981) found a small effect of 100 rad X-irradiation. They irradiated mouse spermatogonia and karyotyped 9- or 10-day-old F_1 fetuses derived from them. The treatment resulted in a threefold increase in abnormalities, including aneuploidy, but the increase in aneuploid frequency alone was not statistically significant.

It could be, as suggested by Russell and Montgomery (1974), that the failure to detect an increase in XO progeny after irradiation may reflect a failure of XO and YO cells to survive in the testis. Evans *et al.* (1969) found that XO spermatocytes in a 39,X/41,XYY mosaic mouse did not occur in the testis and suggested that the Y chromosome was essential for a cell's development in the testis.

Should spermatogonial aneuploidy occur to any appreciable extent and should the aneuploid products contribute to the meiotic pool, this would be most important in terms of long-term genetic effects. The question of whether aneuploidy can be induced through exposure of spermatogonia to an inducing agent is therefore an important one to which we shall return in Chapters 6 and 7.

A more direct measurement of the relative contribution of pre-meiotic errors to meiotic aneuploidy in mammals can be made cytogenetically from the frequency of aneuploidy observed at metaphase I. A note of caution should be added here, however. Most workers do not record metaphase I aneuploidy, and so there is a real possibility that the frequency recorded in the literature is an underestimate. Luthardt *et al.* (1973) reported five hyperploid metaphase I preparations in 1584 mouse oocytes, whilst Speed (1977) recorded two hyperploid mouse spermatocytes in a sample of 318. Dr E. P. Evans has kindly provided additional unpublished data. Of 1895 MI oocytes, 11 had 20 bivalents and one extra body. These data were accumulated, however, over a number of years, and it is not certain that the extra body was a whole chromosome in every case. The corresponding frequency of hypoploid cells cannot be estimated because of the relatively high probability of chromosome loss during the preparative procedures.

THE ANALYSIS OF ANEUPLOIDY ARISING IN ANEUPLOIDS (SECONDARY NONDISJUNCTION)

Whilst the presence of the Y chromosome in XXY females of *Drosophila* had no consequences on the sex determination mechanism, Bridges (1916) showed that it did have an effect on the frequency of sex chromosome aneuploids in the next generation. Females of normal XX constitution typically gave about 0.06 per cent aneuploidy, whilst XXY females gave about 5 per cent sex chromosome exceptions. Bridges (1916) called the process giving rise to aneuploids in XXY females 'secondary nondisjunction' and the aneuploid progeny themselves 'secondary exceptions'. Although Bridges' suggestion that secondary nondisjunction arose from pairing of the X and Y chromosomes, proved later to be inconsistent with subsequent observations (see Gershenson 1933; Sturtevant and Beadle 1936; Grell 1962*a*), there was no doubting the validity of Bridges' conclusion that the increase in aneuploidy was due to the presence of the Y chromosomes in the XXY females, which interfered in some way with the normal disjunction of the X chromosomes.

Bridges had been led to the suggestion that secondary exceptions arose from synapsis of the X and Y chromosomes because the X chromosomes present in female secondary exceptions were invariably non-crossover. However, the presence of the Y chromosome did not bring about a general reduction in the level of crossing-over between the Xs. There was no demonstrable negative correlation between the frequency of secondary nondisjunction and the level of crossing-over which would have been expected had the Y chromosome competed with the Xs for pairing (Anderson 1929; Grell 1962*a*). This observation led Grell (1962*b*) to suggest that pairing and crossing-over occurred prior to a second pairing phase which determined the disjunction of the X and Y chromosomes. On this view of female *Drosophila* meiosis there are two pairing stages; one a necessary precondition of crossing-over, i.e. 'exchange pairing' and a second determining the disjunction of non-crossover chromosomes, i.e. 'distributive pairing'.

There were two observations which were basic foundation stones to the 'distributive pairing hypothesis'. The first was the discovery that extra chromosomes, whilst inducing aneuploidy, were non-competitive in their relationship to exchange. The second was an extensive series of experiments by several workers which revealed the existence of association or pairing of non-homologous chromosomes.

The existence of non-homologous association is inferred from the non-random assortment of non-homologous chromosomes. For example, Grell (1959) observed the non-random assortment of the Y and fourth chromosome and inferred that they segregated away from each other in up to 92 per cent of meioses. The essential feature of the stocks which permitted this demonstration was that one of the fourth chromosomes was translocated onto the third. The remaining 'free' fourth chromosome was marked with a dominant mutation so that its segregation could be followed. In XXY females Grell showed that the Y and the free fourth chromosomes segregated from each other with a high frequency as though they had been associated in meiosis.

Non-random assortment was not confined to the Y and fourth chromosomes. Further studies, for example by Forbes (1960), Ramel (1962), Grell (1963, 1970), Baldwin and Chovnick (1967), and Holm and Chovnick (1975), showed that, in appropriately constructed stocks, non-homologous association could be demonstrated between any pair of non-homologous chromosomes in the *Drosophila* set. A detailed reference list to this work can be found in Grell (1976).

The genetic method of analysis is at its most powerful when combined with other independent methods. Genetically derived hypotheses when confirmed in this way are most compelling. On the other hand, when unsupported, they are much less readily accepted, no matter how convincing the genetic data. From the very dawn of the genetic era acceptance of genetic theories has always been greatly facilitated by parallel molecular or cytological studies. So it is with the distributive pairing hypothesis. The major objection to the hypothesis as an explanation of the phenomenon of non-homologous association is the absence of independent cytological confirmation of the postulated pairing phase. White (1973) has commented that the hypothesis is entirely a genetic one, and Novitski and Puro (1978) have given detailed criticisms of the claims (Grell 1967; Grell and Day 1970; Moore 1971) that in mitotic metaphase, there is an attraction between non-homologous chromosomes. Such observations had been regarded as cytological confirmation of distributive pairing.

Novitski (1964) has observed that the largely non-competitive interaction of extra elements with cross-over bivalents does not necessarily imply that non-homologous association occurs after exchange. He has suggested that the genetic data are also compatible with non-homologous association before exchange. According to this idea synapsis of chromosomes removes homologous elements from this association which persists for any remaining non-synapsed chromosome. The persistent association of non-homologues later affects their disjunction at anaphase I. This hypothesis is a direct development of an earlier suggestion by Oksala (1958) that the centromeres of meiotic chromosomes are associated in a chromocentre, and recent work has provided cytological evidence for its existence (Dävring and Sunner 1973, 1976, 1977; Nokkalo and Puro 1976; Puro and Nokkalo 1977). The important feature of this work is not only the demonstration that the chromocentre exists, but that it persists. It has been shown by both groups of workers that the association is maintained until metaphase I and is therefore quite capable of influencing the segregation of non-homologous chromosomes. This is important, because extensive evidence exists for non-homologous association in many different species, with little evidence that the association influences subsequent segregation. For example, in maize McClintock (1933) demonstrated the existence of non-homologous association which, however, did not persist beyond diakinesis, and cytogenetic tests on trisomic and double trisomic maize revealed no evidence for the sort of non-homologous disjunction found in *Drosophila* (Weber 1969). A similar conclusion was reached by Michel and Burnham (1969) who reported a low frequency of non-homologous pairing which, in double trisomic maize, occasionally persisted until metaphase I. However, only a small proportion of the resulting aneuploidy could be accounted for by this pairing. Similarly in the mouse, non-homologous associations have been reported (Hsu *et al.* 1971) but Cattanach (1967*a*) could find no evidence for the non-homologous disjunction of a translocated chromosome and the single X of an XO female. It could well be that in this respect, the genetic observations made in the *Drosophila* female are peculiar to it and not necessarily representative of meiosis in other organisms.

THE RELATIVE FREQUENCY OF THE DIFFERENT CELLULAR DEFECTS

There is, then, good evidence for the existence of a variety of defects at the cellular level, any of which can generate aneuploidy. With the exception of extra replication of chromosomes, for which definitive evidence that it can generate aneuploidy is still needed, each type of defect has been confirmed by direct cytogenetic observations on the chromosomes. What can be said about the relative frequency of each type of error in contributing to the total amount of aneuploidy of meiotic origin? Of the screening methods used to monitor aneuploid frequency, none permits the unambiguous determination of the precise defect. As a consequence it is surprisingly difficult to decide how much aneuploidy is attributable to each defect.

The *Drosophila* data, particularly on aneuploidy involving the sex chromosomes, are certainly most extensive. In this organism the following conclusion can be drawn:

1. Most, if not all, aneuploidy originates at meiosis. Brood analysis shows that there are no clusters of aneuploids, which would be detectable if mitotic nondisjunction made any significant contribution to meiotic aneuploidy (Anderson 1929).

2. Amongst aneuploids arising spontaneously, those arising at the second meiotic division are very infrequent (less than 1/75 000; Merriam and Frost 1964).

3. A substantial proportion of aneuploids arise through nonconjunction as judged by the increased frequency with which non-exchange chromosomes are detected in aneuploids (Merriam and Frost 1964). However, in a *Drosophila* with structurally normal chromosomes the majority of non-crossover bivalents disjoin correctly.

4. There remains a sizeable fraction of aneuploids which contain recombinant chromosomes and which cannot therefore arise through nonconjunction, although, as Merriam and Frost (1964) have pointed out, other possible types of first division error (nondisjunction, premature centromere division, and extra replication) cannot be distinguished from each other.

5. The frequency with which non-recombinant chromosomes contribute to aneuploidy is increased by the presence of extra chromosomes (secondary nondisjunction; Bridges 1916).

6. Cellular events leading specifically to chromosome loss might occur but these are difficult to distinguish from post-meiotic factors such as spermatogenic failure, which might operate selectively to eliminate disomic, but not nullisomic, products.

In *Sordaria* and *Neurospora*, where complete octads can be analysed, there have been fewer experiments but the following conclusions can be drawn:

1. Mitotic nondisjunction, if it occurs, does not lead to clusters of meiotic aneuploids. However, the known instability of the hyperploid nucleus during mitosis makes it difficult to decide whether clusters would be detectable in any case, so the real contribution of mitotic nondisjunction to meiotic aneuploidy is not known.

2. The frequency of aneuploid spores in both *Neurospora* and *Sordaria* varies widely from cross to cross (Griffiths 1979; Bond and McMillan 1979). This, coupled with the fact that, for some types of error, only a fraction of the total number of events is detected, makes it difficult to make definitive statements about the relative frequency of each type of error. Table 3.1 contains a summary of some representative crosses from *Sordaria* and *Neurospora* and illustrates the relative frequency of each observed class. In the table the various ascus types are classified according to their most probable origin.

3. Nonconjunction and nondisjunction have not been distinguished in either *Sordaria* or *Neurospora* because the experiments with multiply marked chromosomes, equivalent to those of Merriam and Frost in *Drosophila*, have not been carried out. It would be interesting to obtain such data from an organism like *Neurospora*, to determine to what extent non-recombinant chromosomes contribute to aneuploidy in an organism other than *Drosophila*.

4. Second division meiotic errors are more frequent than in *Drosophila* especially in *Neurospora* where the *pan* locus used to detect the aneuploids is closely linked to the centromere so that only a fraction of the total second division errors are detected.

One more thing needs to be said. In this chapter the term nondisjunction has been used in the strict sense as defined by Bridges (1913). Nondisjunction is also used in a more general way to indicate any defect of the cell division process giving rise to aneuploidy. This dual usage has arisen because there is no alternative term which is neutral with respect to the origin of aneuploidy. In the remainder of this book the term nondisjunction will often be used in this general sense.

TABLE 3.1.   *The origins of aneuploidy at meiosis as determined by ascus analysis in* Neurospora *and* Sordaria

Organism	Frequency of asci containing aneuploid spores ($\times 10^{-4}$)	Linkage group	Number of aneuploids arising from:					Reference
			Nondisjunction or nonconjunction	Premature centromere division:		Nondisjunction at second division or premature centromere division	Extra replication of the chromosomes or mitotic nondisjunction	
				of both homologues in bivalent	of one homologue in bivalent			
Neurospora	33.3	VI	28	6	39	3*	†	Threlkeld and Stoltz (1970)
Sordaria	8.2	II	58	not detectable	7	10*	15	Fulton (unpublished)
	5.5	II	149	not detectable	14	11*	25	Bond and McMillan (unpublished)
	7.5	IV	67	not detectable	13	16*	not detectable‡	Fulton (unpublished)

* In *Neurospora* the frequency of aneuploids arising at the second meiotic division can be unambiguously determined. In *Sordaria*, because of the spindle overlap, this is not so.
† In *Neurospora*, only the most common classes of asci containing aneuploid spores were recorded.
‡ Because asci containing aneuploid spores are confounded with recombinants in this case.

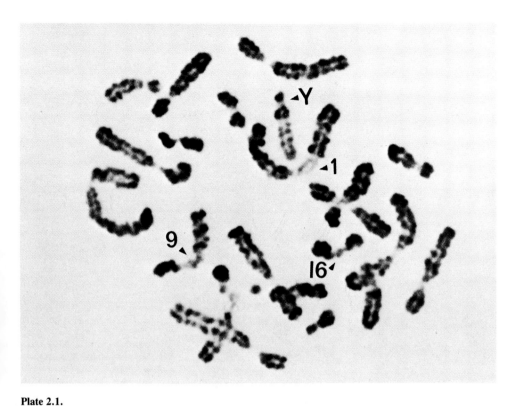

Plate 2.1.
The chromosomes of a human spermatozoon with a 23,Y complement. The heterochromatic regions of chromosomes 1, 9, 16, and the Y are discernible. (Reprinted by permission from *Nature* Vol. 274, pp. 911–13. Copyright 1978, Macmillan Journals Limited.)

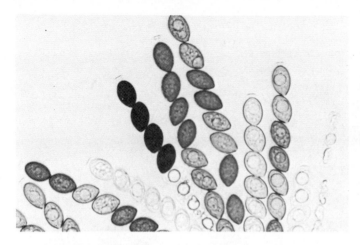

Plate 3.1.
Aneuploid meiotic products in a cross of two complementing *buff* mutants of *Sordaria brevicollis*. Normal asci from this cross contain eight buff ascospores. One ascus is present which arose either from nonconjunction or nondisjunction at the first meiotic division, it contains four disomic, black spores and four nullisomic, abortive, spores. The four disomic spores show full pigmentation because the buff mutants complement each other.

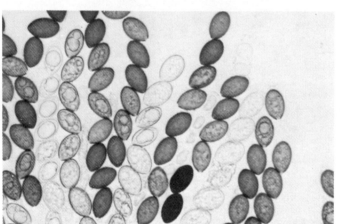

Plate 3.2.
Aneuploid meiotic products arising from premature centromere division in *Sordaria brevicollis*. The ascus containing the two disomic and two nullisomic spores also contains normal haploid spores. The sequence of the spores is important, the fact that the aneuploid spores are in opposite halves of the ascus indicates that they arose from a first-division meiotic error.

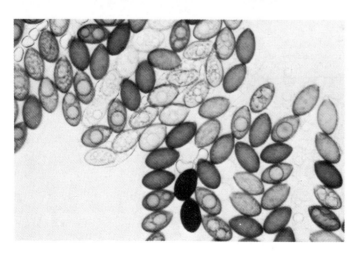

Plate 3.3.
Aneuploid meiotic products possibly arising from nondisjunction at the second division in *Sordaria brevicollis*. The disomic and nullisomic spores are located in the same half of the ascus (cf. Plate 3.2) which is consistent with a second-division origin, but the possibility of spindle overlap which switches the position of the centrally located spores, makes this conclusion uncertain.

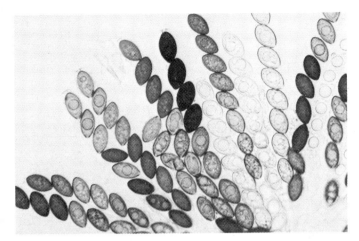

(a)

Plate 3.4.
Disomic spores arising in
meiosis without concommitant
production of nullisomic
spores in *Sordaria brevicollis*.
(a) Ascus with four disomic
black spores and four haploid
buff spores. (b) Ascus with two
disomic black spores and six
buff spores, two of which are
probably disomic buff spores.

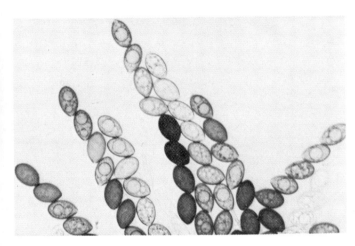

(b)

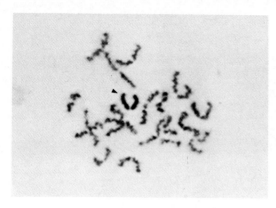

Plate 5.1.
Metaphase II preparations ($n = 20$) from the male mouse (a) X-bearing; (b) Y-bearing. The sex chromosome is arrowed in each case. Preparations stained by conventional means using carbol fuchsin. (c) 'C'-banded preparation.

(a)

(b)

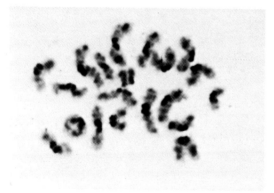

(c)

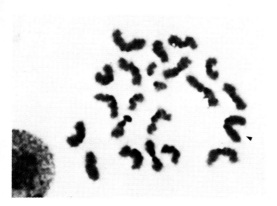

Plate 5.2.

A disomic ($n = 21$) X-bearing metaphase II complement from the male mouse. The X chromosome is arrowed.

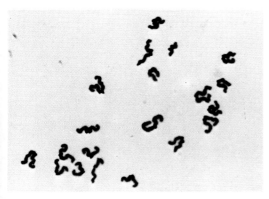

Plate 5.3.

Metaphase II preparations from the female mouse (a) normal complement ($n = 20$); (b) nullisomic complement ($n = 19$), (c) disomic complement ($n = 21$).

(a)

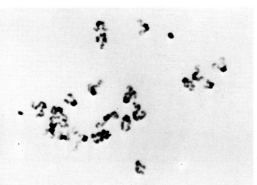

(b)

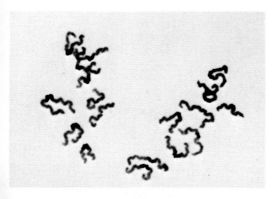

(c)

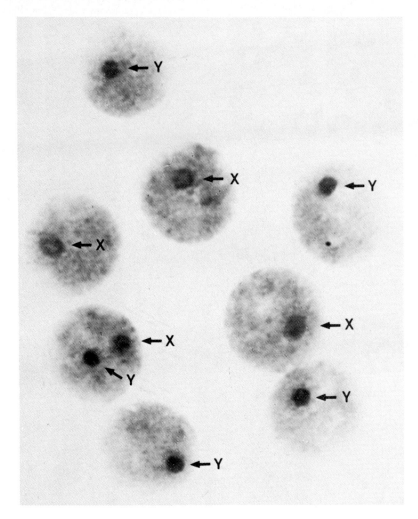

Plate 5.4.
Cyst with nuclei of eight early spermatids from the field vole, *Microtus oeconomus*.
Note the presence of cells with X and Y bodies. One spermatid has an X and a Y body
resulting from sex chromosome nondisjunction. (From Tates (1979). *Envir. Hlth
Perspect.* **31**, 151–9.)

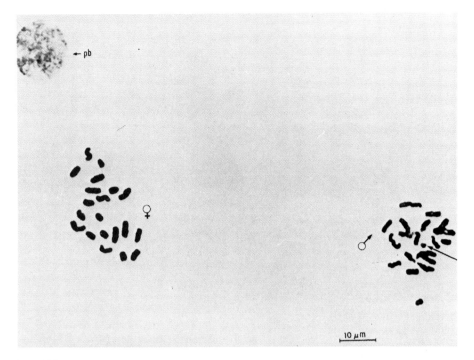

Plate 5.5.
The chromosomes in a first-cleavage mouse embryo. The paternal set can be distinguished from the maternal by its relatively less condensed state of contraction. The small T6 marker chromosome (arrowed) also serves to indicate parental origin. (From Maudlin, I. and Fraser, L. R. (1977). *J. Reprod. Fert.* **50**, 275–80.)

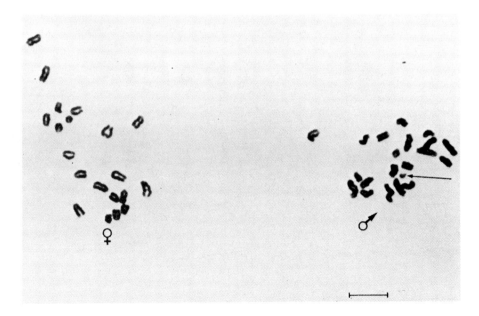

Plate 5.6.
A trisomic chromosome complement from a CBA/H-T6 spermatozoon in a first-cleavage mouse embryo. The T6 marker chromosome is arrowed. (From Maudlin, I. and Fraser, L. R. (1978). *J. Reprod. Fert.* **52**, 107–12.)

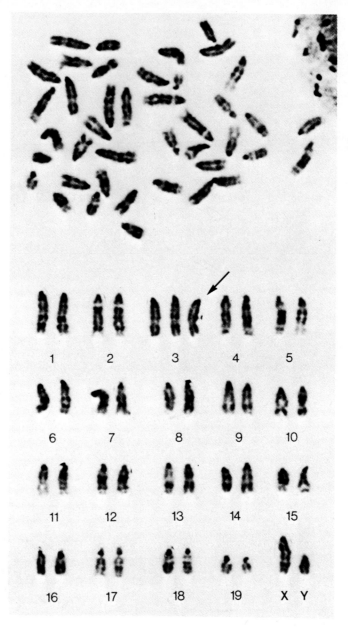

Plate 5.7.
G-banded karyotype of a trisomy 3 male mouse fetus. (From Speed, R. M. and Chandley, A. C. (1981). *Mut. Res.* **84**, 409–18.)

4. The origins and causes of aneuploidy in man

From an analysis of the *Drosophila* and fungal data, we have seen that there is more than one cellular pathway to aneuploidy. How much is known, however, of the cellular defect or defects giving rise to aneuploidy in man? If more than one type of error is involved, what can be said with regard to the frequency of each particular type of defect.

Man is not the most convenient of experimental organisms for a number of obvious reasons. The karyotype does, however, show a variety of polymorphisms which have greatly aided in the tracing of monosomic and trisomic complements, both in liveborn and aborted progeny, back to parental origin. In many cases, the particular meiotic division at which the error arose has been pinpointed. The present chapter gives a detailed description of the results obtained by these types of study. In addition, the various aetiological factors which have been implicated in human aneuploidy induction are outlined and consideration is given to the possible factors underlying the well-known maternal-age effect in man.

THE ORIGIN OF HUMAN TRISOMICS AND MONOSOMICS

Once the various human aneuploid conditions had been identified in the population, interest turned to the possible mechanisms involved in their production and to the relative roles played by maternal and paternal nondisjunction. In the early days, discussion was concerned principally with sex chromosome aneuploidy and determination of the origin of chromosomal imbalance was based on the inheritance of red–green colour blindness (Stern 1959, 1960; Stewart 1960). This was a useful X-linked character because of the frequency of its occurrence and the ease with which it could be detected. For example, if a colour-blind XO female had a colour-blind father and a mother who, by pedigree studies could be safely assumed not to be a carrier of red–green colour blindness, the only X chromosome which the XO woman possessed would have to have been of paternal origin and therefore maternal nondisjunction could be postulated to have occurred. Conversely, if the father of a colour-blind XO woman had normal colour vision, and the mother could be shown to be a carrier, it is obvious that the single X chromosome must have been maternal in origin, and paternal nondisjunction could be inferred. Such studies revealed evidence in some pedigrees of maternal nondisjunction in the genesis of Klinefelter syndrome (47,XXY) and of the maternal origin of the single X chromosome in Turner syndrome (45,X) (Polani 1961).

The discovery of the X-linked dominant blood group gene Xg by Mann *et al.* in 1962 opened the way to further investigations along these lines by providing another simple means whereby the allocation of X-chromosome nondisjunction in families could be determined. Owing to the extensive work of Race and Sanger and their colleagues at the MRC Blood Group Unit in London, a catalogue of the Xg groups of over 2000 patients with various sex chromosome abnormalities, together with their parents, is now available (Race and Sanger 1969; Sanger *et al.* 1971, 1977). For propositi with a 47,XXY karyotype, the extra X chromosome was found by those authors to be paternal (X^mX^PY) in 33 per cent of cases and maternal (X^mX^mY), in 67 per cent of cases, the majority of the errors arising at the first meiotic division (Fig. 4.1). For the 45,X condition, i.e. women with Turner syndrome, Sanger *et al.* (1977) have estimated that 77 per cent of propositae have a maternal (X^mO) and 23 per cent a paternal (X^PO) X chromosome. An excess of maternally derived X chromosomes has also been demonstrated in XO mice, most of these having been thought to arise by paternal sex chromosome loss after sperm entry into the egg. Spontaneously occurring OX^P mice, on the other hand, may arise as the result of maternal meiotic nondisjunction (Russell and Montgomery 1974). Their frequency of occurrence is considerably lower than the X^mO frequency and of the same order of magnitude as the frequency of X^mX^PY offspring which could only arise by nondisjunction at the paternal first meiotic division (Fig. 4.1). The mechanism of origin of the human XO condition still,

SEX CHROMOSOME NONDISJUNCTION
AT   MEIOSIS
IN OOGENESIS AND SPERMATOGENESIS

Normal disjunction

		Normal sperm	
		X^P	Y
Normal egg	X^m	X^mX^P	X^m Y

Maternal nondisjunction

		Normal sperm	
		X^P	Y
ND egg 1st or	X^mX^m	$X^mX^mX^P$	X^mX^mY
2nd division	O	X^PO	YO lethal

Paternal nondisjunction

		N.D. sperm				
		1st division		2nd division		
		X^PY	O	X^PX^P	YY	O
Normal egg	X^m	X^mX^PY	X^mO	$X^mX^PX^P$	X^mYY	X^mO

Fig. 4.1. Sex chromosome nondisjunction at meiosis in oogenesis and spermatogenesis.

however, remains unknown. An explanation for the greater frequency of X^mO compared to OX^P might be that loss of the Y chromosome from XY zygotes is more likely to occur than loss of either X from XX zygotes. From the direct genome analysis of human spermatozoa (Martin *et al*. 1982), it would appear that XY nondisjunction or sex chromosome loss at meiosis in the male do not contribute to any significant extent to the production of XO fetuses. Neither would it appear that maternal meiotic nondisjunction is the main mechanism involved in monosomy-X in man. Recent studies (Kajii and Ohama 1979; Carothers *et al*. 1980; Warburton *et al*. 1980*a*) show that maternal age is not raised in XO conceptuses. On the contrary, the incidence of XO conceptions appears to be highest among young women, a finding which might suggest either that there is an increase in events leading to early cleavage errors in younger women, or perhaps that a greater proportion of their XO conceptions survive to a stage of becoming recognizable pregnancies (Warburton *et al*. 1980*a*).

For the triple-X condition (47,XXX), the data of Sanger *et al*. (1977) for 80 propositae are not clear-cut, but nondisjunction at the first meiotic division in oogenesis seems likely to be the mechanism involved. In the case of the 47,XYY condition, only one meiotic origin is possible, namely paternal nondisjunction at the second meiotic division (Fig. 4.1). However, some cases may be attributable to Y chromosome nondisjunction in a very early division of the XY zygote with subsequent loss of the reciprocal XO line so produced.

Let us now turn to the question of the origins of some of the human autosomal aneuploid conditions. It is because of developments in cytogenetic technique over the past decade that significant advances in this area have also been made. Although the origin of the extra chromosome in autosomal aneuploid conditions like trisomy 21 (Down syndrome) could not be investigated cytologically for a long time, it had nevertheless been assumed that at least some cases originated from malsegregation during oogenesis because of the well known maternal-age dependency (Penrose 1933) of the condition. Already in the pre-banding era, a chromosomal variant giving extremely short arms to a No. 21 chromosome (Gp^-) had been used successfully to identify the maternal origin of the extra chromosome in two patients with Down syndrome (de Grouchy 1970; Juberg and Jones 1970), but it was not until after the improvement of banding techniques (Paris Conference 1971) that the real technical breakthrough came. Soon after the introduction of quinacrine staining to produce fluorescent bands (Q-bands) on human metaphase chromosomes (Caspersson *et al*. 1971) fluorescent polymorphisms were observed, and their potential in tracing the origins of aneuploids and polyploids was quickly realized. Such polymorphisms include alterations in size and/or staining properties of heterochromatic regions and chromosomal satellites and their stalks, and appear to exist without phenotypic effect in the individual carrying them. They are also very stable features of a given chromosome or chromosomal lineage (Schnedl 1971). As the great majority of such heteromorphic regions are situated at or near centromeres, they are virtually unaffected by crossing-over (Mikkelsen *et al.* 1980), and are therefore ideal markers for tracing the origin of chromosome anomalies (Caspersson *et al.* 1970).

The first authors to apply the technique successfully were Licznerski and Lindsten (1972) who traced the origin of the extra chromosome in a trisomy 21 child to its mother. Since then, it has been used not only to determine the mechanism of origin of many autosomal aneuploids among both liveborns and abortuses, but also to trace the origin of polyploids and some *de novo* structural rearrangements (see Jacobs and Hassold 1980, for review). Since some of the most frequently occurring trisomics in spontaneous abortions involve chromosomes which have very well defined polymorphisms, e.g. chromosome 16 and the D- and G-group acrocentrics, the parental origin of the errors can be traced with comparative ease. Observations on the parents and conceptions of 124 spontaneously aborted trisomics from three independent surveys (Lauritsen and Friedrich 1976; Niikawa *et al.* 1977; Hassold and Matsuyama 1979) have been summarized by Jacobs and Hassold (1980) and are presented in Table 4.1. The parental origin was determined in 49 of these 124 cases (40 per cent) and the mechanism involved determined in 39 cases. Irrespective of the chromosome involved in the trisomy, the extra chromosome was almost always maternal in origin, and this held true whether or not the trisomy was strongly maternal age-dependent. Furthermore, in the vast majority of cases, the error occurred at the first meiotic division, errors at the second division being equally infrequent in both males and females.

TABLE 4.1.   *Origin of trisomy—spontaneous abortions (124 cases examined—origin determined 40 per cent)*

Origin				Chromosome						Total	
♂ I	♂ II	♀ I	♀ II	13	14	15	16	21	22		
+	–	–	–				2			2	
–	+	–	–	1			1			2	8%
+	+	–	–							0	
–	–	+	–	3	1	2	12	4	11	33	
–	–	–	+				1	1		2	92%
–	–	+	+	1		2	1	3	3	10	
				5	1	4	17	8	14	49	

From: Jacobs, P. A. and Hassold, T. J. (1980). In *Human embryonic and fetal death* (ed. I. H. Porter and E. B. Hook). Academic Press, New York.

TABLE 4.2.   *Origin of liveborn trisomy 21 (368 cases examined—origin determined 46 per cent)*

Origin				Fully reported		Partially reported	
♂ I	♂ II	♀ I	♀ II				
+	–	–	–	0		16	
–	+	–	–	2	8%	16	24%
+	+	–	–	0		2	
–	–	+	–	14		88	
–	–	–	+	3	92%	17	76%
–	–	+	+	5		5	
				24		144	

From: Jacobs, P. A. and Hassold, T. J. (1980). In *Human embryonic and fetal death* (ed. I. H. Porter and E. B. Hook). Academic Press, New York.

Data from Jacobs and Hassold (1980) concerning the origin of the additional chromosome in liveborn trisomy 21 cases are given in Table 4.2. Here the findings differ somewhat from those in spontaneous abortions. A total of 368 cases have been examined and the parental origin of the extra chromosome established in 168 (46 per cent). As in the spontaneous abortion series, nondisjunction at the first meiotic division in oogenesis is by far the most common cause of the extra chromosome in these liveborn cases of trisomy 21, accounting for over 60 per cent of cases. However, there appears to be a substantial increase in both paternal and second-division errors among liveborn trisomics compared with those found in abortions. Some of the liveborn surveys for example give almost identical frequencies for maternal and paternal origins of the extra chromosome 21 (Wagenbichler *et al.* 1976) whilst others give figures indicating a 20–30 per cent paternal contribution (Mikkelsen *et al.* 1976; Mattei *et al.* 1979). Others, however, have found the paternal contribution to be less than 10 per cent, as in abortions (Robinson 1973; Magenis *et al.* 1977).

The reasons for the differences between various surveys are not clear but may be due, at least in part, to the manner in which the data have been reported (Hassold and Matsuyama 1979). With the exception of Robinson's study (Robinson 1973), none of the reports of liveborn individuals gave information on all cases studied. In contrast, of the three studies using spontaneous abortions, only Lauritsen and Friedrich (1976) reported just those cases in which a determination of the parental origin had been made. Such incomplete reporting results in loss of information but more importantly, by considering only cases informative for a specific mechanism of origin, a bias towards second-division errors is introduced (Robinson 1973, 1977; Langenbeck *et al.* 1976; Jacobs and Morton 1977).

The fact that most of the studies of liveborn individuals have been biased in this manner suggests that the frequency of second-division errors may well be artificially inflated in the liveborns relative to the abortion series (Hassold and Matsuyama 1979). Alternatively, the presence of undetected mosaicism in liveborn trisomics may increase the frequency of apparent second-division errors (Warburton *et al.* 1978), since trisomics resulting from post-zygotic nondisjunction are indistinguishable cytologically from second-division errors (Fig. 4.2). If, therefore, such mosaic trisomic conceptuses had a better chance of surviving to term than did complete trisomics, the effect of mosaicism on the frequency of apparent second-division errors would be more pronounced in data from liveborns than from abortuses.

It is not quite so obvious, however, why incomplete reporting or hidden mosaicism should artificially inflate the numbers of errors due to paternal nondisjunction. Mikkelsen *et al.* (1980) have suggested in this connection that the disparity between the abortion data and the liveborn data may possibly reflect different parental age distributions.

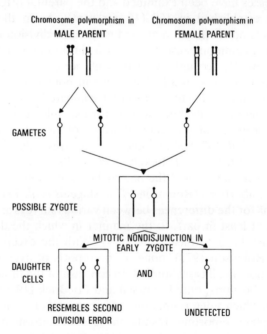

Fig. 4.2.   A diagram to show how a mitotic nondisjunction event in an early zygote can resemble a second division error in one of the parents if the mosaicism resulting from the mitotic event is undetected.

In their most recent study, tracing the origins of the extra chromosome in trisomy 21 liveborns (Mikkelsen *et al.* 1980), various combined staining techniques, e.g. Q-banding plus silver staining of the acrocentric nucleolar organizer regions (Goodpasture and Bloom 1975; Mikelsaar *et al.* 1977) have been applied with great success. The extra No. 21 chromosome in liveborn Down individuals has been identified by them successfully in 76 per cent of cases, a success rate nearly double that found in previous studies (Mikkelsen *et al.* 1976; Jacobs and Hassold 1980). In their 1980 study, Mikkelsen and her colleagues made comparisons for two Danish populations, one on the rural island of Funen, the other on the island of Zealand on which is situated Copenhagen. The origin of the extra chromosome was determined as maternal or paternal in 77 per cent of the 60 trisomic probands from Funen and 76 per cent of the 45 trisomic probands from Zealand (Table 4.3). In both populations, maternal meiotic first-division failure was the most common origin, accounting for 61 per cent of the informative cases, a figure in excellent agreement with that of Jacobs and Hassold (1980). Second meiotic failure in the mother was found in 20 per cent of cases from Funen, but only 9 per cent of cases from Zealand. Paternal nondisjunction was observed in 11 per cent of cases from Funen (three first-division and two second-division errors), a figure in accord with that of Jacobs and Hassold (1980) for Hawaii, but in 23.5 per cent of cases (all first-division errors) from Zealand (Table 4.3). This high level of paternal nondisjunction was not significantly different from that found previously by Mikkelsen *et al.* (1976) for individuals from Copenhagen. It was, however, significantly higher than the 11 per cent of paternal errors recorded for the Funen population. Since the maternal-age distribution was lower on Zealand, Mikkelsen *et al.* (1980) considered that

TABLE 4.3. *Type of nondisjunction for liveborn patients with trisomy 21 from the Danish islands of Funen and Zealand*

Type of meiotic error	Funen No.	Zealand No.	Total No.
Maternal I	28	21	49
Maternal II	9	3	12
Maternal ?	3	2	5
Paternal I	3	7	10
Paternal II	2	0	2
Paternal ?	0	1	1
Unknown	14	11	25
Mitotic error (crossing over)	1	0	1
	60	45	105

From: Mikkelsen *et al.* (1980).

this could perhaps explain the higher relative paternal contribution on that island. Environmental factors operating in the urban areas could, however, she believed, also be playing a part in promoting paternal nondisjunction.

Finally, it is clear from the data concerning spontaneously aborted (Hassold and Matsuyama 1979; Hassold *et al.* 1980*a,b*) and liveborn (Magenis *et al.* 1977; Jacobs and Hassold 1980; Mikkelsen *et al.* 1976, 1980) trisomics that maternal first meiotic division errors are the most common cause of the trisomy, and this holds true both at high and low maternal ages. The factors which could play a role in influencing events at meiosis in general, and at the first maternal meiotic division in particular, will be the subject for discussion in a later part of this chapter.

THE AGE EFFECT

In 1876, Fraser and Mitchell first drew attention to the fact that children with Down syndrome tended to be born at the end of large sibships. Penrose (1933, 1934, 1954) was able to show that this increase of Down syndrome births was related to the age of the mother and not to any other single factor such as paternal age or parity. The fact that the incidence of Down syndrome increases with maternal age is now, of course, well documented and the 'maternal-age effect' has been detected across national and racial categories, with a remarkable uniformity in age-independent incidence rates (Penrose 1933; Lilienfeld 1969). At about the

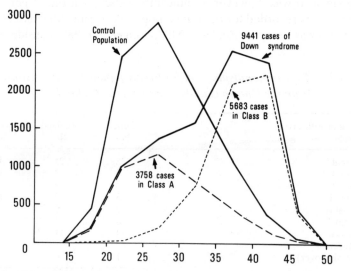

Fig. 4.3.   Maternal age distribution of 9441 cases of Down syndrome with control population. (From Smith, G. F. and Berg, J. M. (1976). *Down's anomaly*, 2nd edn. Churchill Livingstone, Edinburgh, with permission.)

maternal age of 20 years, the incidence is 1 in 2300 births, at ages 30–34 about 1 in 880, and at about 45 years and over, it increases markedly to about 1 in 54 (Smith and Berg 1976).

Down syndrome typically shows a bimodal maternal age distribution (see Fig. 4.3), and Penrose and Smith (1966) have suggested that this could best be accounted for by assuming two components, one age-independent (Class A), the other age-dependent (Class B). The maternal-age-independent cases show a peak frequency of Down syndrome births at 28.5 years, corresponding to the peak for all births in the population, while the age-dependent class shows a peak around 43 years. Class A Down syndrome comprise about two-fifths of all affected subjects and includes all hereditary cases (translocation trisomics) while the remaining three-fifths (Class B), (primary trisomics), owe their genesis to some process associated with increasing age of the mother (Penrose and Smith 1966). There is also a maternal-age effect for liveborn trisomy 13 and 18 individuals (Magenis *et al.* 1968; Taylor 1968), the other two autosomal trisomic conditions which occur with substantial frequency in liveborns. Liveborn trisomy 8 subjects, however, appear to show little, if any, relationship to increasing age of the mother (Riccardi 1977).

A maternal-age effect has also been found for many trisomic abortuses. In a recent extensive study carried out by Hassold *et al.* (1980*a*) the effect of maternal age on the genesis of trisomy was studied by comparing data from 362 trisomic and 790 chromosomally normal spontaneous abortions. As a group, the trisomics were associated with a substantial increase in maternal age, but there were considerable differences in the magnitude of the effect for different trisomics. The effect was most pronounced for trisomics involving the small chromosomes, whether acrocentric or non-acrocentric. However, trisomy 16 was conspicuously different from all other small chromosome trisomics, both in the reduced importance of maternal age and in the high frequency with which it occurred (see Chapter 2). Trisomy for chromosomes in groups A, B and C also was associated with only a moderate increase in maternal age.

There are now also suggestions in the literature of an increased incidence of Down syndrome at very low maternal ages. Erickson (1978) has presented data (Fig. 4.4) showing that the incidence rates for mothers aged 15 years or less (based, however, on only 15 cases) are equivalent to those of mothers 30–35 years of age. More extensive data are required to confirm whether or not this is a real effect.

In recent years, there have been several reports from around the world, that the mean maternal age in Down syndrome is declining (Collman and Stoller 1969; Uchida 1970; Jones and Lowry 1975; Mikkelsen *et al.* 1976; Nordensen 1979; Koulischer and Gillerot 1980). Uchida (1970), for example, noted that the incidence of Down syndrome in a Manitoba population was constant over many years at about 1 in 600, as in the rest of

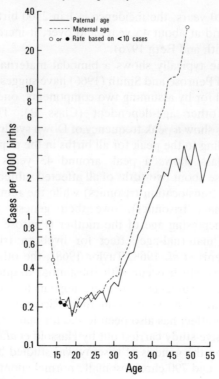

Fig. 4.4. Down syndrome rates by parental age, NIS surveillance areas 1961–6. (From Erickson, J.D. (1978). *Ann. hum. Genet.* **41**, 289–98, with permission.)

the world. However, the mean maternal age of this population fell from 27.2 years in 1960 to 26.2 years in 1967, through a decrease in the number of births to women in the 30–39 age group and an increase among those aged 15–24 years. From this, Uchida concluded that the incidence of Down syndrome in the Manitoba population should have fallen, and she suggested that new environmental factors, in addition to maternal age, might be operating to keep it at 1 in 600. In Denmark too, Mikkelsen *et al.* (1976) found an unchanged incidence of Down syndrome throughout a twelve-year period (1960–1971), even though the maternal-age distribution had shown a shift to younger ages. Similar observations have been made in Sweden (Nordensen 1979) and Belgium (Koulischer and Gillerot 1980). The possibility that paternal as well as maternal nondisjunction errors are becoming more common, particularly in heavily urbanized areas, has been considered by some authors as an explanation for these observations (Mikkelsen *et al.* 1976, 1980; Nordensen 1979; Uchida 1979). Koulischer and Gillerot (1980), however, have put forward an alternative suggestion. They compared the two four-year periods 1971–1974 and 1975–1978, and recorded an increased frequency of Down syndrome in women below 35 years from Wallonia in Belgium. During the second period, this reached a level of statistical significance in the Charleroi region, an industrial and densely populated area. The authors looked for new factors operating in Wallonia limited to mothers below 35 years and acting particularly in the Charleroi region. They noted the close coincidence in time of their observations on Down syndrome births with the legalization in Belgium of information on contraception, and with an increase in contraceptive users. They believe that careless and erratic use of the contraceptive pill with resulting 'pregnancy breakthrough' might lie behind some of the extra aneuploidy encountered among the younger mothers in this region. Evidence for increased levels of spontaneous abortion after oral contraceptive failure has been presented in the literature over the last few years (e.g. Harlap *et al.* 1979, 1980).

In the past, a *paternal* age effect for Down syndrome has been difficult to establish, but not everyone has completely accepted the idea of a sole effect of age of mother. Mantel and Stark (1967) for example pointed out that paternal age may also have an effect, but because of the inevitable high correlation within marriages between maternal and paternal ages, any search for it could be rendered ineffectual since such correlations would make the sizes of the samples used effectively equivalent to much smaller ones. The recent knowledge gained from cytogenetic investigations into the origins of the extra chromosome in trisomy 21 subjects where, at least for liveborns, 10–20 per cent of cases may be paternal in origin (Mikkelsen *et al.* 1980), has, however, aroused renewed interest in the possibility of a paternal-age effect. Stene *et al.* (1977) have considered the deficiencies of the statistical methods which were used in the past for separating the effects of maternal and paternal age, and seemingly demonstrated, using a more powerful approach, a paternal age effect in a Danish sample of Down syndrome infants. Moreover, from prenatal diagnosis information, obtained in West Germany, these authors (Stene *et al.* 1981) recently have shown the paternal-age effect for men of 41 years and upwards to be quite strong. The risk for a fetus to have any *de novo* chromosomal aberration increased more with advancing paternal age for older mothers than for young ones. These authors believe, therefore, that the ages of both parents should always be taken into account as an indication for prenatal diagnosis. Erickson (1978), however, in a statistical study of the associaton between Down syndrome incidence, maternal age, paternal age, and birth order in over 4000 cases, failed to find any independent effect of paternal age. In fact, the rates at paternal ages over 45 years appeared to be nearly constant.

For the sex chromosomal aneuploidies, there have been several reports of increased maternal age at birth compared with control groups (Lenz *et al.* 1959; Ferguson-Smith *et al.* 1964; Court Brown *et al.* 1969; Borgaonkar and Mules 1970; Tumba 1974; Carothers *et al.* 1978). Court Brown *et al.* (1969) for example, found evidence for a maternal-age-dependent component in XXY males, XXX females, and double aneuploids, e.g. XXY+G, XXY+E, XY, +E, +G, etc. but no significant deviation from the control maternal-age distribution for XO females or XYY males. In a more recent study, which included some of the old data used in the study made by Court Brown *et al.* (1969), Carothers *et al.* (1978) showed that maternal age was the most likely causative factor in the XXY condition, while birth order was the least likely for the XXXs. A small but significant inverse relationship between paternal age and the incidence of XYYs (exclusively paternal in meiotic origin) was also found. The validity of the method used in the analysis was confirmed by applying it to a group of familial chromosome rearrangements which, as expected, revealed no dependence on parental age or birth order. A common feature of all these studies, however, is that propositi are ascertained, not through random samples of the general population, but through selected categories such as those attending endocrine, subfertility or psychiatric clinics, inmates of hospitals for the mentally subnormal and so on. An alternative approach, which avoids the possibility of bias inherent in such ascertainments, is based on surveys of the consecutive newborn. Unfortunately, the number of aneuploids detected in this way is necessarily rather small so that the results of several such surveys must be pooled in order to obtain useful results. In just such an exercise, Robinson *et al.* (1979) found evidence of a raised maternal age for XXX births, but not for XXY or XYY ones. The discrepancy between the two approaches in the XXY results may simply be a small-sample effect. Alternatively, since the mean year of birth of the propositi from the newborn surveys is necessarily more recent than that of the propositi in the non-random studies, it may be due to an increasing incidence in recent years of non-age-related XXY births, similar to that suggested earlier for Down syndrome. Further data are needed to resolve this point.

Carothers *et al.* (1980) have also studied the data for liveborn propositae with Turner syndrome (45,X) and have confirmed earlier studies (e.g. Ferguson-Smith 1965; Court Brown *et al.* 1969; Borgaonkar and Mules 1970) in showing no positive association with parental age or birth order: in fact, the findings suggest a small negative association. Kajii and Ohama (1979) and Warburton *et al.* (1980*a*) have also reported an inverse maternal age effect for XO abortuses. The reduced maternal ages noted among XO abortuses in these two studies indicates that the main mechanism responsible for the occurrence of monosomy X is not nondisjunction during maternal meiosis. It is assumed that at least some XO conceptions are due to chromosome loss following anaphase lagging at meiosis or mitosis and that this is, for some reason, more common among younger women (Kajii and Ohama 1979). Warburton *et al.* (1980*a*) have suggested alternatively that an increased rate of survival of XO conceptuses to the stage of producing recognizable pregnancies could be occurring in younger women. Since the same maternal-age relationship for monosomy X has now been observed in New York, Geneva, Honolulu, and Hiroshima, populations very different in racial and cultural make-up, any possible environmental factor would have to be widespread throughout the world (Warburton *et al.* 1980*a*).

In somatic cells too, it would appear that aging can influence mitotic aneuploidy induction in man. Aneuploidy of the X chromosome in the blood lymphocytes of older subjects has been well documented (Jacobs *et al.* 1963; Galloway and Buckton 1978), the frequency of aneuploid cells showing a significant increase with age in both males and females, the effect being, however, more marked in females. The loss of an X chromosome (leading to XO) is much more common than a gain (leading to XXX). In females, the occurrence of a 'fragment' of an X-chromosome also correlates with increasing age (Fitzgerald *et al.* 1975; Galloway and Buckton 1978), and this fragment appears to be an X chromosome that has simply divided prematurely at the centromere. Such premature division of the centromere occurs *in vivo* and leads to acquired sex chromosome aneuploidy in cells of aging women, but the increases become obvious only in women who are past child-bearing age (Galloway and Buckton 1978). In older males, loss of the Y chromosome was reported to occur by Jacobs *et al.* (1963), but this observation could not be confirmed by Galloway and Buckton (1978). In both sexes, cells lacking a G-group acrocentric chromosome make a significant contribution to the total in all studies.

AETIOLOGICAL FACTORS IN HUMAN ANEUPLOIDY

Among all the aspects of human aneuploidy which have been considered, none has received more attention than the aetiological factors which might play a role in their production. Yet on this subject alone there is probably a greater amount of equivocal data than on any other. The fact is that we are really not very much nearer today to pinning down the responsible mechanisms than we were twenty years ago when the human aneuploid conditions were first identified.

A clear distinction which should be made when considering the aetiology of aneuploidy is that different factors may be at play in producing the age-dependent errors, compared with those which are responsible for the age-independent anomalies. When a particular trisomic condition is known to be maternal-age-dependent, then the most likely cause of its induction is one which is directly or indirectly related to the aging of the female or the ovum itself, although other causative factors are not excluded. For trisomic conditions which are maternal-age-independent, however, other causative factors, such as paternal nondisjunction, genetic predisposition to nondisjunction or exogenous forces may come more into play. (The hypothesis that nondisjunction occurs at an equal rate in all maternal age groups but that rejection of trisomic conceptuses decreases with maternal aging, receives little or no support at this time (Smith and Berg 1976).) The fact that first meiotic maternal errors predominate both in strongly age-related (e.g. trisomy 21) and weakly age-related (e.g. trisomy 16) conditions (Hassold *et al.* 1980*a*) as well as at all maternal ages (Hassold and Matsuyama 1979; Mikkelsen *et al.* 1980), suggests, indeed, that there could be at least two different kinds of mechanism operating to produce the errors, one of which is dependent on, the other independent of, or only minimally influenced by maternal age.

One early suggestion to explain age-dependent trisomy 21 was put forward by Penrose in 1965. According to his hypothesis, the age effect could arise because of the accumulation of a series of errors leading eventually to nondisjunction. These errors associated with aging could represent decay of a limited and irreplaceable number of chromosomal spindle fibres with a limited half-life. Their gradual disappearance would lead to chromosome traction imbalance at first meiotic division and to consequent nondisjunction. The idea that the spindle fibres may decay in the long female meiotic prophase is attractive, but there seems to be no evidence that they are present throughout dictyotene (or diplotene). Spindle tubules or other components may, however, be formed during meiotic prophase, in advance of the assembly of the fibres which form the spindle proper, active at the meiotic division, and they could be subject to wear and tear over the many years which can elapse in women before ovulation (Alberman *et al.* 1972). It is also possible that noxious environmental factors like X-rays could accelerate their decay (Alberman *et al.* 1972). Experimental evidence that nondisjunction at the first meiotic division is associated with defects in the meiotic spindle fibres in eggs aged by preovulatory overripeness has in fact been found in studies with *Xenopus* (Mikamo 1968), and the rat (Butcher and Fugo 1967; Mikamo and Hamaguchi 1975; Kamiguchi *et al.* 1979).

Another hypothesis which has been put forward to explain the increasing frequency of human trisomy 21 with age of the mother is that delayed fertilization of a normally ovulated egg might lead to nondisjunction (German 1968). In several non-human species the incidence of aneuploid and mosaic as well as polyploid, embryos has been found to be increased after experimentally induced delayed fertilization (Witschi and Laguens 1963; Austin 1967; Butcher and Fugo 1967; Vickers 1969; Yamamoto and Ingalls 1972). This has been attributed to a disturbance of chromosomal segregation in the egg suspended in metaphase II for a few hours too long before a spermatozoon becomes available to penetrate it and cause syngamy. The anomalies are believed to be produced by disintegration of the second meiotic spindle in aging eggs (Yamamoto and Ingalls 1972;

Sugawara and Mikamo 1980). In the mouse, Rodman (1971) has in fact observed that aged secondary oocytes at metaphase II may manifest a fragmentation of chromatids at their centromeric junctions and this could account for some aneuploids having their genesis at the second meiotic division. There is no evidence that metaphase II of oogenesis is any less vulnerable in man than in the other mammals examined so far, and German (1968) has postulated that the maternal-age effect in women could be a reflection of decreasing frequency of coitus with increasing duration of marriage, thereby delaying the time of fertilization of the ovum in older women. German's statistical data give support to his hypothesis, but a subsequent study carried out by Penrose and Berg (1968) does not. Moreover, Cannings and Cannings (1968) have shown that the age dependency for the frequency of coitus is not closely enough related to the incidence of Down syndrome to support German's theory. Data collected from a Baltimore study (Sigler *et al.* 1965) also fail to provide support for the hypothesis (Lilienfeld 1969). Other authors have criticized German's hypothesis on different grounds. James (1978) for example, has argued that if trisomy 21 were directly caused by low coital rates, then one might expect the births of Down syndrome infants to be preceded by unusually long fallow periods and the fertility preceding such births to be lower than that preceding the birth of normal infants (controlling for maternal age). Neither expectation is fulfilled. Also, if it were true that delayed fertilization could increase nondisjunction, the effect should only be seen at the second meiotic division of the egg, and a preponderance of second meiotic errors would therefore be expected among trisomics born to older mothers. There are no indications that this is the case, the majority of errors, as stated before, being chiefly ones which have arisen at the first meiotic division of meiosis, irrespective of maternal age (Mikkelsen *et al.* 1980). Nevertheless, in spite of these objections, the basic tenet of the German hypothesis has been invoked to explain an observed doubling in the incidence of Down syndrome births among Catholic women in Western Australia compared to women in all other religious groups (Mulcahy 1978). Here, the ovulatory method of birth control, in which timing of intercourse is discordant with ovulation, is the suggested mechanism leading to delayed fertilization. Similar increases in the incidence of Down syndrome births among Catholic women as compared to women in other religious groups have been noted in The Netherlands (Jongbloet *et al.* 1978).

To explain the maternal age dependancy of trisomy 21, it has also been argued that nondisjunction may result from persistence of the nucleolus at meiotic prophase leading to failure of pairing in the small acrocentric pairs in aged oocytes (Polani *et al.* 1960). Moreover, as the nucleolus is present in the resting oocyte, and persists to diakinesis or first meiotic metaphase, failure of its dissolution could lead to nondisjunction of the nucleolus-associated D- and G-group acrocentrics at anaphase I. This effect could be enhanced by nucleotropic viral infections (Evans 1967). Recent cytological observations made at pachytene in the human oocyte (Mirre *et al.* 1980) show the sites of ribosomal genes on as many as three nucleolar (acrocentric) bivalents juxtaposed inside a nucleolar fibrillar centre, one of whose major components is a silver-positive stained protein. Mirre *et al.* (1980) suggest that this protein would normally disappear when the oocyte undergoes the first meiotic division, but in aged oocytes, some enzyme deficiency may cause it to persist, thus hindering chromosome segregation in the acrocentric pairs. As Hassold and Matsuyama (1979) have pointed out, however, the occurrence of both age-related trisomics (e.g. trisomy 18) and age-independent trisomics (e.g. trisomy 16) involving chromosomes *without* nucleolar organizing regions would limit the general applicability of this explanation. Hassold *et al.* (1980*a*) have, in fact, found age-related increases in trisomy for the great majority of the small chromosomes (i.e. Nos. 13–22) among human spontaneous abortions, both acrocentric *and* non-acrocentric. In fact, the three highest mean maternal ages were seen in trisomics involving non-acrocentric chromosomes, namely those of chromosomes 17, 18, and 20. Thus, in view of those authors, it is unwarranted to attribute the effect of increasing maternal age to factors such as 'retention of nucleoli', which affect only acrocentric chromosomes.

Some evidence for nucleolar persistence has been found, however, in somatic cells showing satellite associations between acrocentric chromosomes (Ferguson-Smith and Handmaker 1961). A significantly higher frequency of such associations has been found for the No. 21 chromosome pair in the mothers of children with Down syndrome by some authors (Hansson and Mikkelsen 1978; Hansson 1979) but not by others (Cooke and Curtis 1974; Davison *et al.* 1981). Such increases have also been reported in the lymphocytes of women using oral contraceptives (McQuarrie *et al.* 1970). The idea that high satellite association may increase the risk of nondisjunction is supported by the finding in Hansson and Mikkelsen's study that the satellite association tendency of chromosome 21 in the parent responsible for the nondisjunctional error was significantly higher than in controls. The relation between satellite association and age, however, appears equivocal (Hansson 1979). Robinson and Newton (1977) have shown that there is an increased occurrence of a particular kind of polymorphism, namely that called 'positive satellites', among Down syndrome individuals compared with controls, and that the No. 21s with such intense or brilliant satellites associate significantly less often than do those with normal satellites. Robinson and Newton (1977) suggest that there may be an association between polymorphic types and tendency to undergo nondisjunction which, if confirmed, might provide a means whereby a woman's risk of producing a Down child could be estimated with considerably more accuracy than is possible from age alone.

Yet another suggestion which has been put forward to explain the maternal-age effect is that of Henderson and Edwards (1968) who found declining chiasma frequencies and increasing numbers of univalents in aging mouse oocytes. They suggested, in their 'production line' hypothesis, that random segregation of such univalents at metaphase I could account for the excess of trisomics born to older mice and, perhaps, older women. Declining chiasma frequencies have been claimed for aged human oocytes (Luthardt 1977) but generally speaking the poor quality of fixation in human oocytes cultured to metaphase I of meiosis, makes accurate interpretation of chiasma positions and frequency extremely difficult, if not impossible.

The original observation of a declining chiasma frequency and/or increasing univalent frequency with age, made, in the mouse, by Henderson and Edwards (1968), has since been confirmed for that species by other authors (Luthardt *et al.* 1973; Polani and Jagiello 1976; Speed 1977; de Boer and van der Hoeven 1980), and has been reported too for the Chinese hamster (Sugawara and Mikamo 1983). Neither Polani and Jagiello (1976) nor Speed (1977), however, could confirm the expected association between an increased occurrence of univalents at metaphase I and hyperploidy at metaphase II in the mouse. This indicates either that these univalents can undergo regular segregation, or that oocytes containing univalents fail to progress through the first meiotic division. Strong and convincing evidence for the former idea comes from the recent work of Sugawara and Mikamo (1983) using the Chinese hamster. They have demonstrated quite clearly that no correlation exists, in that species, between the pairs of chromosomes seen as univalents in MI oocytes (which always belong to the smallest size group) and those undergoing nondisjunction (which belong to all size groups). Univalent presence at MI does not therefore necessarily lead to aneuploidy at MII. Indeed, even Polani and Jagiello (1976), who have consistently argued in favour of Henderson and Edwards' 'production line' hypothesis, have speculated that the univalent pairs seen in their MI oocytes of the mouse could have been artefactual and attributable to preparative technique.

Henderson and Edwards (1968) proposed the 'production line' hypothesis to account for their observations of lower chiasma frequencies and higher numbers of univalents in the MI oocytes of old mice. They suggested that cells entering meiosis late in fetal life might form fewer chiasmata (and thus more univalents) than those entering early, and that they would be ovulated later in reproductive life. Gradients (developmental or nutritional) were postulated to occur in the fetal ovary to account for the decline in chiasma frequency.

Now, if age-related nondisjunction does relate back to events in the fetal ovary, and if a production line exists, then it might be possible, by cytological means, to detect an increase in synaptic irregularities and/or univalents, or a declining chiasma frequency in oocytes over gestational time. In fact, some studies have been carried out to this end. Jagiello and Fang (1979), for example, analysed the diplotene oocytes observed on days 16 and 18 of gestation in Swiss mice and claimed to find a decline in numbers of chiasmata. Speed and Chandley (in press), using the same strain, were, however, unable to confirm these findings. They could not find diplotene oocytes on day 16, and in those seen on day 18, the chromosomal elements were not of sufficient clarity to allow for analysis. They, moreover, were unable to find any other cytological evidence of a production line, either in the mouse or human ovary (Speed, in preparation).

Speed and Chandley (in press) believe, as do Sugawara and Mikamo (1983), that it is some factor in the adult maternal environment which is the real cause of age-related aneuploidy. The latter authors suggested that increasing levels of nondisjunction with maternal age could relate to a higher incidence of spindle defects in old oocytes. de Boer and van der Hoeven (1980) have suggested that 'poorer physiological conditions' in the adult ovarian follicles of older females might be responsible. Hormonal imbalance towards the end of reproductive life might also play a part. It is known that adequate steroid support is necessary for normal oocyte development in the follicle (Moor 1978). The concept of a hormonal factor in the aetiology of Down syndrome was proposed by Rundle *et al.* (1961) in early studies showing that significantly higher levels of dihydro-epiandrosterone were present in the urine of women who had borne a Down syndrome child when young compared with those producing such a child late in life. Lyon and Hawker (1973) have also proposed that age-related changes in hormone levels could bring about disturbances in meiotic chromosome segregation in mice and a hypothesis suggesting an interaction between hormones and meiotic segregation has recently also been proposed by Crowley *et al.* (1979) to explain the maternal age effect for trisomy 21 in man. Any hormonal imbalance which occurred at the beginning or end of a woman's reproductive life, might account for the observed increases in nondisjunction during these two periods. Menstrual cycle length irregularities, with associated hormonal fluctuations, do indeed occur in women during the five to seven years succeeding menarche and the six to eight years preceding the menopause (Treloar *et al.* 1967; Sherman *et al.* 1976). It may be relevant that in mice, the strain (CBA) in which the highest numbers of univalents and the most significant decline in chiasma frequency were found by Henderson and Edwards (1968) is also a strain showing a strong maternal-age effect (Gosden 1973; Fabricant and Schneider 1978; Brook 1983), and a short reproductive span in which oocyte depletion is complete at the end of reproductive life (Jones and Krohn 1961). In human females too, the ovary at the menopause is almost completely devoid of oocytes (Costoff and Mahesh 1975), and it has been suggested by Noyes (1970), that more chromosomally defective ova might be ovulated in a species or strain showing such complete usage of its oocytes. There would be less likelihood for this to occur in a strain in which many eggs remained in the ovary beyond the end of the fertile period.

A further implication for the role of hormones or hormonal imbalance in human aneuploidy production comes from studies on women who have had ovulation-inducing therapy in a cycle before or during which fertilization occurred (Boué and Boué 1973*b*; Boué *et al.* 1975; Alberman 1978) or those who have had pregnancy breakthrough while on the pill (Koulischer and Gillerot 1980; Harlap *et al.* 1979, 1980). These situations have all given indications for raised frequencies of aneuploid births or first-trimester abortions. Further data are required, however, both epidemiological and experimental, before a certain link between hormones and nondisjunction can be established with certainty (see also p. 143).

Epidemiological studies indicate that X-rays may be a causative factor in nondisjunction both in women (see Wald *et al.* 1970 and Uchida 1979 for review) and men (Boué *et al.* 1975) although the data are somewhat equivocal (see Chapter 6). One epidemiological study has also suggested a relationship between Down syndrome and paternal exposure to radar (Cohen and Lilienfeld 1970) and importance has been attached by some workers to high fluoride content of water (Rapaport 1963) and to atmospheric pollution (Greenberg 1964).

Several epidemiological studies on operating room personnel have indicated relatively high frequencies of abortion (Cohen *et al.* 1971; Knill-Jones 1975) and/or congenital abnormality (Corbett *et al.* 1974; Pharoah *et al.* 1977) among their progeny and these effects may, at least in part, be due to aneuploidy induction. The effects are mainly attributed to the inhalation of waste anaesthetic gases, in particular halothane.

An increased incidence of thyroid antibodies has been reported in mothers of children with Down syndrome, but its significance is not clear (Fialkow *et al.* 1965). An increased incidence of thyroid disease (McDonald 1972) or a faulty immune mechanism (Zsako and Kaplan 1969) may also be important. Also, a recent investigation into parental α_1-antitrypsin (PI) types suggests that PI deficiency may cause interference with some mechanism during cell division thus leading to nondisjunction. A striking fivefold increase in the frequency of the 'MS' and 'MZ' types of PI variants (which are recognized as deficiency variants) was found in mothers where nondisjunction leading to Down syndrome had occurred during the first meiotic division (Jongbloet *et al.* 1981).

Significant cyclic seasonal patterns of aneuploidy superimposed on the age-dependent incidence rates of trisomy 21 have been reported from Australia (Collman and Stoller 1962), Canada (Zarfas and Wolf 1979), Israel (Wahrman and Fried 1970), Sweden (Lander *et al.* 1964), and the

United States (Goad *et al.* 1976) and seasonal variation in aneuploidy for the sex chromosomes has been reported by Nielsen and Friedrich (1969), Jongbloet (1971), and Bell and Corey (1974). However, other authors (e.g. Edwards 1961; Froland 1967; Stark and Mantel 1967; Carothers *et al.* 1980) report no significant seasonal effects. Carothers *et al.* (1980) have reviewed the conflicting data from the literature on seasonal variation in chromosomal aberrations and have urged caution in the interpretation of the results. They have suggested that the existence of a seasonal association with the birth incidence of an abnormality points to an environmental cause, but that when dealing with conditions with a high prenatal mortality it should be borne in mind that the environmental effect may be one that influences not origin, but prenatal viability.

Secondary or 'inevitable' nondisjunction in a fully affected or mosaic fertile trisomic individual can of course also give rise to trisomic offspring as also can segregation from a Dq Gq or Gq Gq translocation heterozygote. Approximately 5 per cent of all Down syndrome individuals are 'translocation trisomics' arising in this way. The risk of trisomy 21 has been found also to be five to 10 times higher in women carrying a Dq Dq Robertsonian translocation (Dutrillaux and Lejeune 1970) and about 20 times higher in the sibs of trisomics for chromosome 18 (Hecht *et al.* 1964). These and other co-occurrences within individuals or families have been accounted for by Grell and Valencia (1964) and Grell (1971*a*) by reference to the 'distributive pairing hypothesis' first put forward to explain similar events in *Drosophila* (see Chapter 3). These authors believe that the presence of rearranged or extra chromosomes in human kinships and the occurrence of aneuploidy for an unrelated chromosome in the same kinship, are probably not fortuitous events.

It has also been proposed that chromosomal polymorphisms of constitutive heterochromatin on certain chromosomes, e.g. pairs No. 1, 9, and 16 may predispose them to undergo nondisjunction (Geraedts and Pearson 1973; Nielsen *et al.* 1974; Hassold 1980*a*). Nielsen *et al.* (1974) suggested that duplication of the heterochromatic region in chromosome 9 might affect pairing and segregation thus giving an increased risk of nondisjunction. Other anomalies such as elongated short arms in an acrocentric chromosome or abnormalities in the satellite region of such chromosomes might also predispose to nondisjunction (Hamerton *et al.* 1965; Davison *et al.* 1981). Alternatively, segregation might be affected by the position of a particular chromosome in relation to other chromosomes in the genome (Rodman *et al.* 1980; Bennett 1982) or by its size and C-band constitution (Ford and Lester 1982). All, or some of these, might be important in determining why certain chromosomes undergo nondisjunction more frequently than others.

In summary, it can be seen from the foregoing, that on the vexed question of aetiology, much has been said and written but very little has actually been achieved in rendering more lucid our understanding of the precise mechanisms involved in the production of human aneuploidy. Probably a combination of factors is at work. Exogenous factors for example may be superimposed on natural aging processes while certain chromosomes may show special predisposing features which render them more liable to undergo irregular segregation.

Current emphasis in much of the work on nondisjunction in mammals is therefore centred on the experimental induction of aneuploidy, and comparisons with lower organisms, in which a good deal of data are now available, are being attempted. The experimental approach in mammals, however, will be deferred until later in the book (see Chapters 6 and 7).

THE POSSIBLE EXISTENCE OF MEIOTIC MUTANTS PROMOTING NONDISJUNCTION IN MAN

The finding of strain differences in the age-dependency of nondisjunction in oocytes of mice (e.g. Martin *et al.* 1976; Fabricant and Schneider 1978) suggests that spontaneous aneuploidy levels may be genetically determined. This, in turn, could be attributable to genetically controlled factors such as hormonal control of the ovary (Jones and Krohn 1961), but it does point to the possibility that in man too, predisposition to nondisjunction in different individuals or races could occur. Several studies dealing with this aspect of the aneuploidy question have been reviewed (Hook and Porter 1977), but it is these authors' belief that there is no definite evidence in the literature for consistent differences between races in the prevalence, at birth, at least of Down syndrome.

In individual women and within pedigrees or families, there are, however, indications that mutations affecting nondisjunction might exist. Several groups of authors have observed a tendency for trisomic karyotypes to recur among consecutive spontaneous abortions of the same women (Boué and Boué 1973*c*; Hassold 1980; Alberman 1981). Moreover, several different trisomic conditions are sometimes encountered in various closely related family members (Baikie *et al.* 1961; Miller *et al.* 1961; Dumon and Leroy 1980; Alfi *et al.* 1980). This apparently genetically determined faulty control of chromosome movement might, at times, affect several chromosome pairs and the several trisomic conditions which are encountered could thus be explained. Some even believe that an association may exist between repeated spontaneous abortion and parental sex chromosome mosaicism (Hecht 1982). The latter could therefore be used as an indicator of an increased risk of meiotic nondisjunction. As Baker *et al.* (1976) have pointed out, however, it is by no means clear to what extent such non-randomness is caused by meiotic mutants in man. Other possibilities, for which some evidence exists, are physiological, viral, or other environmental factors predisposing to anomalous chromosome behaviour. It is also possible that non-homologous chromosomes can segregate from one another in man as they do in *Drosophila* (Grell and Valencia 1964; Grell 1971*a*) and there are the obvious technical problems of undetected translocations, parental gonadal mosaicism, and ascertainment bias. The existence of mutants which control chromosome segregation in man has yet to be proven.

5. Spontaneous aneuploidy in mammals other than man

OCCURRENCE OF VIABLE ANEUPLOID TYPES

Although a large part of our attention has so far been focused on aneuploidy in man, it should not be forgotten that aneuploid individuals are also sporadically found in other mammalian species. Since large chromosome surveys of laboratory and domestic animals are rarely performed, the discovery of a viable aneuploid animal generally comes about through some phenotypic abnormality, notably sterility. Others have, however, been found by use of appropriate genetic markers in experiments specially designed for the purpose of detecting viable aneuploid individuals (see below). A list of reported spontaneous occurrences of sex chromosome aneuploidy among mammals is given in Table 5.1. For the autosomal aneuploids, there is scant information since it is supposed that, as in man, the majority would not survive to full term. A sterile male mouse, thought to be trisomic for chromosome 16, was,

TABLE 5.1. *Sex chromosome aneuploidy in mammals other than man*

Anomaly	Species	Reference
XO	mouse	Welshons and Russell (1959)
	pig	Nes (1968)
	rhesus monkey	Weiss *et al.* (1973)
	sheep	Zartman *et al.* (1981)
	cat	Norby *et al.* (1974)
	horse	Chandley *et al.* (1975)
	wallaby	Sharman *et al.* (1970)
	black rat	Yong (1971)
	mole rat	Sharma and Raman (1971)
	field mouse	Bianchi and Contreras (1967)
XXX	cow	Rieck *et al.* (1970)
	horse	Chandley *et al.* (1975)
XXY	mouse	Cattanach (1961)
		Russell and Chu (1961)
	cat	Centerwell and Benirschke (1973)
	dog	Clough *et al.* (1970)
	sheep	Bruère *et al.* (1969)
	ox	Rieck (1970)
	pig	Breeuwsma (1968)
	wallaby	Sharman *et al.* (1970)
XYY	mouse	Cattanach and Pollard (1969)

however, reported by Cattanach (1964). Spontaneous aneuploidy, among karyotyped mouse fetuses, is not an uncommon type of chromosome abnormality (see below) and it is also possible to generate aneuploids by means of a special breeding system which permits the induction of specific aneuploid conditions (Gropp *et al.* 1975). Using one parent doubly heterozygous for two partially homologous Robertsonian metacentrics, several monosomic and all nineteen trisomic conditions in the mouse have been produced using this approach. From such studies, it is known that certain monosomics are eliminated before implantation while others die shortly after (Gropp and Epstein 1982). Some trisomics do not survive a first critical phase of organogenesis on days 11 to 12 of fetal development, others (Ts 12 – 14, 16, 18 and 19) have a life-span until or beyond birth. Many questions related to the developmental pathology of fetal aneuploidy can thus be investigated systematically using this mouse model system.

Cytogenetic screening for the detection of aneuploidy in laboratory and domestic animals, although not carried out to any great extent among adult animals, has been carried out on germ cells, one-cell embryos and mid-term fetuses. Genetic methods have also been developed for the detection of aneuploidy among liveborn individuals. From the outset, the experimenter must decide, therefore, whether he requires a technique for assessing initial levels of aneuploidy in germ cells or early embryos, or whether one which provides a figure for surviving aneuploidy at a subsequent stage of fetal or adult development will suffice.

In this chapter a detailed account of the various methods available for aneuploidy detection in mammals will be given together with the spontaneous levels recorded. In Chapter 6 we will go on to consider aneuploidy induction using physical agents, and in Chapter 7 the use of chemical agents in aneuploidy induction will be described. The literature is extensive and therefore subdivided under a number of headings.

METHODS OF DETECTION OF ANEUPLOIDY

In germ cells

The simplest way to detect aneuploidy arising at the first meiotic division in mammalian germ cells is to count chromosomes at the second meiotic metaphase in secondary oocytes or spermatocytes.

In males, many second meiotic metaphases are usually present on slides prepared from any normal testicular suspension and chromosome spreading in air-dried preparations can generally be obtained which is good enough to allow for unambiguous counting of chromosome arms. Several authors have adopted this procedure for aneuploidy analysis in the male mouse (Beatty *et al.* 1975; Polani and Jagiello 1976), where a chromosome count of $n=20$ is the modal number (Plate 5.1a and b) and counts of $n=19$

and $n=21$ indicate nullisomy and disomy respectively (Plate 5.2). Moreover, the darker staining properties of the X and Y chromosomes at metaphase II in the male mouse (Plate 5.1a and b) allows for their identification, and nondisjunction rates for the sex chromosome pair can be estimated separately from those of the autosomes (Ohno *et al.* 1959). A further refinement comes from the use of an appropriate 'C'-banding method for centromeric staining which can allow unambiguous counting of chromosomes in the absence of really adequate spreading (Hsu *et al.* 1971; Polani 1972) (Plate 5.1c). In females, the analysis of metaphase II chromosomes (Plate 5.3) can be carried out either in eggs removed from the adult ovary at the arrested diplotene or 'dictyate' stage and cultured *in vitro* through the first meiotic division (Edwards 1965; Uchida and Lee 1974; Polani and Jagiello 1976) or alternatively, in oocytes collected from the oviduct following natural or hormonally induced ovulation (Jagiello *et al.* 1968; Röhrborn and Hansmann 1971; Hansmann and Probeck 1979; de Boer and van der Hoeven 1980). The levels of gonadotrophic hormone required to achieve superovulation have been found, in a number of species, not to increase aneuploidy levels over and above spontaneous frequencies (Fechheimer and Beatty 1974; Maudlin and Fraser 1977; Hansmann and El-Nahass 1979; de Boer and van der Hoeven 1980; Golbus 1981). Differences in aneuploidy levels have, however, been reported for oocytes from the same strain prepared by these different methods. For young females of the Swiss and CBA strains, Golbus (1981) found more hyperploid metaphase II counts when eggs were prepared by *in vitro* maturation than when collected by natural or super-ovulation. No such difference was seen, however, when oocytes from old females of these two strains were prepared. The authors' belief was that their *in vitro* culture system, which gave a higher success rate than that reported by other workers, may have allowed the maturation of more 'marginal' oocytes from young, but not old, females. This is obviously a consideration which must be taken into account when data presented by different workers using alternative egg preparation procedures are compared. The fixation of eggs is usually carried out by the standard technique of Tarkowski (1966), but a modified procedure designed to minimize artefactual chromosome loss has been described by Kamiguchi *et al.* (1976). Since chromosome loss due to preparative technique tends artificially to inflate the numbers of nullisomic metaphase II complements, correction is made by many authors when aneuploidy levels are quoted. This entails a doubling of the hyperploid counts, hypoploid counts being ignored (Beatty *et al.* 1975). As in spermatocytes, C-banding results in an enhancement in quality of the oocyte metaphase II chromosomes and provides the means for reliable and unambiguous counting of the chromosomal elements (Plate 5.3).

Analysis of oocytes or spermatocytes at metaphase II will of course provide an estimate of first meiotic nondisjunction. To assess the total levels of nondisjunction arising at *both* the meiotic divisions, analysis must be made in a post-meiotic germ-cell stage or in the first-cleavage embryo. A method to estimate levels of sex chromosome nondisjunction arising at both meiotic divisions in the spermatocyte of the northern vole, *Microtus oeconomus*, has been described by Tates *et al.* (1975) and Tates (1979). In that species, the X and Y chromosomes are large and darkly staining with C-banding and can be identified in the round spermatids as 'X' or 'Y' bodies (Plate 5.4). Two such bodies are seen when nondisjunction in the sex pair has occurred (Plate 5.4). Since, in practice, many hundreds of round spermatids are present on the slides prepared from any male with normal spermatogenesis, a large population of cells can be screened in a very short time. This type of approach has also been tried in man, where it has been claimed that dark-staining or fluorescent bodies in the spermatozoa can be analysed to assess levels of nondisjunction for chromosomes 1, 9, 16, and the Y chromosome (see Chapter 2). These studies have, however, been criticized on a number of grounds and little credence is now attached to them. The more direct approach of analysing human sperm chromosomes following *in vitro* fertilization into zona-free golden hamster eggs (Rudak *et al.* 1978; Martin *et al.* 1982) is now the preferred method (see Chapter 2).

In first-cleavage embryos

An alternative means of assessing aneuploidy levels arising at the two meiotic divisions is to make chromosome counts at the pronuclear stage in the zygote. Here, the chromosomal sets can usually be identified as maternal or paternal by their relative degrees of contraction (Plate 5.5), the sperm-derived chromosomes being less condensed than those of the egg (Donahue 1972; Maudlin and Fraser 1977; Tease 1982). A marker chromosome introduced from one or other parent can be used as an additional check on parental origin (Donahue 1972; Maudlin and Fraser 1977, 1978*a*) (Plate 5.5). An embryo showing a sperm-derived trisomic complement marked with the small T6 chromosome is shown in Plate 5.6. Fertilization of the egg may be carried out by either an *in vivo* (Donahue 1972; Kaufmann 1973; Maudlin and Fraser 1977, 1978*a,b*; Fraser and Maudlin 1979; Tease 1982), or *in vitro* (Maudlin and Fraser 1977, 1978*a,b*; Fraser and Maudlin 1979) technique. Since no significant differences in the incidence of either monosomy or trisomy between embryos fertilized *in vivo* or *in vitro* were detected by Maudlin and Fraser (1977, 1978*a*) it would appear that the system itself does not generate such anomalies.

In fetuses

The detection of aneuploidy at subsequent times in gestation has been carried out in a range of species including mouse, rat, Chinese hamster, rabbit, pig, sheep, and cow (see Table 5.4) In the mouse, where most of the effort has been concentrated, systematic karyotyping has been carried out on morulae and blastocysts (Gosden 1973) and in fetuses over days 8–15 (Yamamoto *et al.* 1973*a*; Ford and Evans 1973; Fabricant and Schneider 1978; Chandley and Speed 1979; Speed and Chandley 1981). The technique employed is usually that of Evans *et al.* (1972). In this species, it is now known that the vast majority of autosomal monosomics are eliminated before implantation, the rest dying shortly thereafter (Gropp and Epstein 1982). For the trisomics, there is good survival until at least days 11 or 12, but then they too succumb to death. The advantage of karyotyping morulae or blastocysts is, therefore, that a greater proportion of aneuploid types can be recovered. The advantage of karyotyping 9–10-day-old fetuses is that a greater number of mitotic divisions are available for analysis on the slides and when a numerical abnormality is detected, precise information regarding the chromosome involved in nondisjunction can be obtained by application of a suitable G-banding technique (Plate 5.7). The relative contributions made by sex-chromosomal and autosomal nondisjunction can then be assessed.

In liveborns

Routine cytogenetic screening of unselected human newborn populations has produced much information on the frequency of various aneuploids at birth in man (see Chapter 2). In other mammals, however, routine karyotyping among newborns has not been carried out to any great extent and genetic testing among adults is hampered by the dearth of good experimental systems. Nevertheless, Goodlin (1965) has karyotyped a large population of newborn mice from aged mothers. Also, sex chromosome aneuploidy levels in adult mice have been studied by observing the atypical inheritance of X-linked marker genes (Russell 1961, 1979; Russell and Saylors 1963) a method based on the fact that nondisjunction which involves the sex chromosomes produces mostly viable mice. Simple nondisjunction of the sex chromosomes can produce the XO, OY, XXX, XXY, and XYY types. Detection of exceptional types results from the circumstances that, in mammals, the genes on only one X chromosome, at random, are active, so that animals carrying two X chromosomes with different markers are mosaics (Russell 1979) (Fig. 5.1).

By means of 'complementation' testing, in which two gametes with complementary aneusomies give rise to chromosomally balanced viable zygotes, autosomal nondisjunction rates can also be assessed (Lyon *et al.* 1976). Such systems employ marker genes in the detection of the nondisjunctional event. For further details of the genetic systems required in such analyses, the reader is, however, referred to the original references cited.

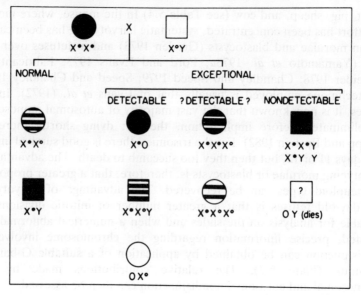

Fig. 5.1. Diagrammatic representation of phenotypes in the simplest mating scheme for the detection of numerical sex chromosome anomalies. Allelic markers on the X chromosome are represented as 'black' ($X^\bullet$) and 'white' (X°). Females are shown by circles and males by squares. Striping indicates mosaicism, with the relative widths of the stripes representing the approximate proportion of the two cell types in the mosaic. (From Russell, L. B. (1979). *Envir. Hlth Perspect.* **31**, 113–22.)

When each of these different endpoints for detecting aneuploidy in mammals is considered, however, none is found to be entirely free from shortcoming or limitation. As Russell (1979) has pointed out, one problem with metaphase II cytology is that the identity of individual chromosomes cannot be adequately established. Also, artefactual loss of a single chromosome can inflate the numbers of hypoploid counts. The scoring of extra 'X' or 'Y' bodies in the spermatids of *Microtus oeconomus* (Tates *et al.* 1975; Tates 1979), although a quick procedure, could detect anomalies additional to whole-chromosome aneuploidy. One disadvantage of the scoring of cells in mid-gestation embryos is that several cells per conceptus must be scored instead of just one, as in metaphase II analysis. Also, when chromosomal mosaics are found, it is often difficult to decide whether they are purely mitotic (post-conception events), or meiotic hyperploidies plus subsequent mitotic loss. Conversely, non-mosaic hyperploidies need not necessarily be due to meiotic events: since the embryo derives from only a very few descendants of the blastomeres, early mitotic nondisjunction does not always result in a mosaic. Another disadvantage of this procedure is that the later the embryo is scored, the smaller is the chance of discovering autosomal aneuploidy. The chance of finding autosomal monosomy is almost nil.

Of the cytological methods available, therefore, scoring in the early cleavage divisions through the blastocyst stage appears to be the most likely one to avoid some of the above difficulties, but even this method is laborious and time-consuming.

The chief disadvantage to the screening of liveborn numerical sex chromosome anomalies is that the diagnostic trisomic type will be relatively rare. On the plus side, however, the genotype of sex chromosome trisomics by this method gives clues as to the mechanism of their origin in a way that cytological analysis cannot accomplish.

Using one or other of the various test systems described above, different groups of workers have assessed both the spontaneous and experimentally induced levels of aneuploidy in a variety of mammalian species. Moreover, studies have been made for different strains within a species, and at different ages within a strain.

SPONTANEOUS ANEUPLOIDY LEVELS IN MAMMALS

In germ cells

Spontaneous nondisjunction of the sex chromosomes at anaphase I in the male mouse appears to be an exceedingly rare event. In an early investigation carried out by Ohno *et al.* (1959), every one of 1460 second meiotic divisions examined from males of six different strains showed a single X or Y chromosome (Plate 5.1). Not one example of abnormal sex chromosome segregation was recorded. Total aneuploidy was recorded by Beatty *et al.* (1975) for thirteen different strains. Altogether, 5200 well-spread C-banded metaphase II spermatocytes were analysed and ten were found to contain 21 chromosomes. Of these, nine contained an extra autosome and one contained two Y chromosomes. The overall level of aneuploidy was 0.38 ± 0.12 per cent, a figure obtained by doubling the number of hyperploid cells to allow for the expected class of hypoploid chromosome complements arising from true nondisjunction and not attributable to artefactual chromosome loss.

Data on metaphase II counts in spermatocytes from male mice of two different strains (CSI and CFLP) and at a range of ages from 0.6–18.5 months were obtained by Polani and Jagiello (1976). Strain differences were not apparent and by pooling together the data for young (0.6–4 months) and old (6–18.5 months) males, the results showed a non-significant difference of 2.08 per cent aneuploidy in young and 1.16 per cent in aged males. Where information is given (for the CSI strain), five out of ten disomic complements had a probable extra sex chromosome and five an extra autosome.

For the northern vole, *Microtus oeconomus*, Tates *et al.* (1979) reported exceedingly low levels of spontaneous nondisjunction for the XY pair in spermatocytes. In one series of experiments, (Tates 1979) only three aneuploid spermatids (2XY and 1YY) were detected among 40 000 cells from sixteen males, while in a second series (Tates and de Vogel 1981), no aneuploids were found among 8000 spermatids from eight males.

For mouse oocytes at the metaphase II stage, more data are available and these are summarized in Table 5.2. Many different strains have been tested, a range of ages has been used and different methods of egg maturation have been employed. The result is a confusing array of results. For simplicity, the data will be pooled over two age ranges namely 2–6 months (young oocytes) and 7–18.5 months (old oocytes). Aneuploidy frequencies will be calculated by doubling the hyperploid counts given by each set of authors.

For the young age range, it is clear that in all the studies, with the exception of those made by Golbus (1981), aneuploidy levels of 1.0 per cent or less are found. Strain differences are not apparent and the chosen method of egg culture would appear not to influence the results. The conflicting results for the Swiss and CBA strains of Golbus (1981) compared with Brook (1983) and Martin *et al.* (1976) over this age range however remain unexplained. The much higher levels of aneuploidy recorded by Golbus (1981), particularly from *in vitro* maturation of young eggs, has been attributed, by that author, to a higher culture success rate compared with that found in other studies.

Within the aged group, levels of aneuploidy from 0.0 to 7.12 per cent have been recorded. As in the young group, however, the levels of aneuploidy recorded by Golbus (1981) far exceed those of all other authors. For those studies in which 'within-strain' comparisons have been made for young and aged oocytes (Uchida and Lee 1974; Polani and Jagiello 1976; Martin *et al.* 1976; Uchida and Freeman 1977; de Boer and van der Hoeven 1980; Golbus 1981), the results are variable. For example, Polani and Jagiello (1976) using the CSI and CFLP strains and Golbus (1981) using the CBA and Swiss strains, could find no age-related increases in aneuploidy over the age ranges tested up to 18.5 months (see Table 5.2). Uchida and Lee (1974) and Uchida and Freeman (1977) did, however, find a small but significant age-related increase in aneuploidy for their (C3H × 1CR/Swiss) F_1 females tested at 3–6 months and at 12 months. Using a special homozygous translocation stock (T(1; 13)70H) in which long and short marker chromosomes could readily be identified in the metaphase II preparations, de Boer and van der Hoeven (1980) recorded a ninefold rise in nondisjunction for the short marker alone (from 0.16 to 1.42 per cent,

TABLE 5.2. *Spontaneous aneuploidy in mouse oocytes, analysed at metaphase II. Results of various independent studies*

Age range in months	Reference	Strain	No. oocytes analysed	Chromosome 20 or less	count >20	Total aneuploidy‡ (%)
1–2	*Golbus (1981)	Swiss	450	439	11	5.02
	†Golbus (1981)	Swiss	350	346	4	2.32
	*Golbus (1981)	CBA	275	258	17	12.36
	†Golbus (1981)	CBA	275	273	2	1.44
2–3	†Hansmann and El Nahass (1979)	C3H/HeHan	1302	1298	4	0.62
		(101 × C3H) F_1	740	737	3	0.82
		NMRI/Han	1296	1293	3	0.46
Young 2.5–3	†Reichert *et al.* (1975)	NMRJ	143	143	0	0.00
2–5	*Martin *et al.* (1976)	CBA	187	187	0	0.00
0–6	*Polani and Jagiello (1976)	CSI	233	232	1	0.86
	Polani and Jagiello (1976)	CFLP	119	119	0	0.00
3–4	†Brook, J. D. (1983)	Swiss	218	218	0	0.00
	de Boer and van der Hoeven (1980)	T (1; 13) 70H homozygotes	645	643	2	0.62
3–6	*Uchida and Lee (1974)	(C3H × 1CR/Swiss) F_1	1054	1054	0	0.00
5–8	*Martin *et al.* (1976)	CBA	116	110	6	10.40
7–18.5	*Polani and Jagiello (1976)	CSI	114	114	0	0.00
	Polani and Jagiello (1976)	CFLP	162	161	1	1.24
8.5–11	*Martin *et al.* (1976)	CBA	155	155	0	0.00
9–10	*Golbus (1981)	CBA	200	194	6	6.00
	†Golbus (1981)	CBA	200	196	4	4.00
Aged 11–14	de Boer and van der Hoeven (1980)	T (1; 13) 70H homozygotes	560	556	4	1.40
12	*Uchida and Freeman (1977)	(C3H × 1CR/Swiss) F_1	1306	1302	4	0.62
	*Speed (1977)	Q	58	57	1	3.44
12–15	*Golbus (1981)	Swiss	225	217	8	7.12
	†Golbus (1981)	Swiss	225	218	7	6.22

* *In vitro* cultured metaphase II eggs.

† Superovulated metaphase II eggs.

‡ Total aneuploidy = twice the hyperploid frequency.

with age, but only a 1.7-fold rise over the same age range (3–4 months to 11–14 months) for the remaining chromosomes. The overall levels of aneuploidy for all chromosomes were 0.62 per cent at young and 1.4 per cent at old age. The failure of Golbus (1981) to detect an age-related increase in CBA oocytes is surprising in view of other studies which have demonstrated a maternal-age effect for aneuploidy in this strain when F_1 embryos have been analysed over a similar age range (Gosden 1973; Fabricant and Schneider 1978) (see Table 5.4). The issue becomes even more complicated when the CBA oocyte data of Martin *et al.* (1976) are considered. In those studies females of the CBA strain showed an overall level of aneuploidy of 10.4 per cent in middle-aged females (5–8 months) compared with no aneuploidy in females which were young (2–5 months) or aged (8.5–11 months). The authors suggested that the lack of hyperploid oocytes in their old CBA females might have been due to a threshold effect in which oocytes containing many univalents at metaphase I became atretic and failed to progress to metaphase II. Fabricant and Schneider (1978) have also reported a plateau effect in age-related aneuploidy in the CBA strain, but in their studies the decline was not seen until after 10 months (see Table 5.4)

It is interesting to note here that Lyon *et al.* (1976), in genetic complementation studies, also recorded an increase in autosomal nondis-junction as female mice carrying Robertsonian translocations passed from young (2–4 months) to mid-aged (5–7 months) but the level then declined in females eight months and older. These authors also suggested that opposing factors were at work, one tending to increase, the other to decrease the observed frequency of marked young.

Data obtained at metaphase II in females of mammalian species other than the mouse are limited but some are available. In the Syrian hamster (*Mesocricetus auratus*) and Chinese hamster (*Cricetulus griseus*), oocytes obtained by superovulation from 8–20-week old animals were examined for chromosomal imbalance by Hansmann and Probeck (1979). Only one hyperploid oocyte among 307 studied was detected in the former and none in 125 in the latter species. Mikamo (1979), also working with the Chinese hamster, reported an incidence of nondisjunction in the female first meiotic division of 1.8 per cent for young females and in a subsequent study Sugawara and Mikamo (1983) recorded levels of 1.5 per cent for 5–8-month-old and 3.6 per cent for 16–18-month-old females of that species. The difference was significant ($p < 0.05$). Furthermore, from an analysis of 1423 1–2-cell embryos, Mikamo (1979) deduced a frequency of 0.9 per cent nondisjunction in the second meiotic division. When aged (16–19 months) females were tested, it was found that the abnormalities which increased with maternal age were mostly aneuploids originating at the first meiotic division.

Spontaneous aneuploidy in early and mid-gestation embryos

From an analysis of first cleavage embryos, the overall levels of aneuploidy arising at both meiotic divisions in the mouse have been estimated and the contribution coming from the male and female parent respectively assessed. In an early study using *in vivo* fertilization, Donahue (1972) recorded a level of 1.8 per cent aneuploidy for the young mouse and Maudlin and Fraser (1977) later provided data in good agreement with that figure (Table 5.3). A slightly lower figure of 0.4 per cent aneuploidy was recorded in one-cell embryos of the 3HI strain by Tease (1982). Maudlin and Fraser's (1977) estimates of 1.9 per cent for *in vitro* and 0.7 per cent for *in vivo* fertilized embryos (a non-significant difference) were based only on those hyperploid chromosome spreads in which no 'irregularities' were seen and which could be scored with confidence. In a later study carried out by Maudlin and Fraser (1978*a*), a frequency of aneuploidy of around 2 per cent was found, the aneuploids being divided equally between male- and female-derived chromosomal complements and between trisomic and monosomic anomalies. From these studies, therefore, it would seem that the recorded levels of about 1 per cent aneuploidy in metaphase II oocytes of young females with a roughly comparable contribution coming from the male, would be consistent with a frequency of about 2 per cent for the early cleavage mouse embryo.

As gestation proceeds, so aneuploid fetuses are resorbed and the frequency of detectable monosomics and trisomics declines. The detected levels of aneuploidy (based on a doubling of all recorded trisomic complements) among morulae, blastocysts, and mid-gestation fetuses are shown in Table 5.3. For morulae and blastocysts of one-month-old CBA mothers, Gosden (1973) recorded a level of 6.0 per cent aneuploidy. For aged (8–12-month-old) mothers of the same strain, the level rose to 21 per cent (Table 5.4), a statistically significant age-related increase. For mid-gestation embryos, the recorded aneuploidy frequencies are generally low, ranging from 2.8 per cent at 6–7 days (Takagi and Sasaki 1976) to 0.0 per cent at 10–14 days (Ford and Evans 1973; Fabricant and Schneider 1978). A declining frequency of recovery of aneuploid fetuses as gestation proceeds is of course anticipated in view of the death of virtually all monosomics before implantation and most autosomal trisomics from about day 10 of gestation onwards (Gropp and Epstein 1982). By birth in the mouse, recorded levels of aneuploidy are thus negligible. The data are limited, but when Goodlin (1965) examined the karyotypes of 842 newborn (BALB/C × 129)F_1 mice, no aneuploids were found amongst them. Similarly, Russell and Saylors (1963) recorded only one XO exception among 3059 offspring classified in a control series of experiments carried out in conjunction with irradiation studies (see Chapter 6).

When comparisons between the frequencies of aneuploid F_1 embryos

TABLE 5.3.   *Incidence of aneuploidy in first cleavage and early- to mid-gestation embryos of the young mouse (Aneuploidy incidence obtained by doubling the hyperploid counts)*

Embryonic stage	Reference	Strain	No. embryos analysed	% aneuploidy
1 cell	*Tease (1982)	3H1	523	0.4
1 cell	*Maudlin and Fraser (1977)	TO	429	0.7
1 cell	†Maudlin and Fraser (1977)	TO	853	1.9
1 cell	*Donahue (1972)	not stated	338	1.8
1 cell	*Kaufman (1973)	CFLP	193	0.5
Morulae and blastocysts	Gosden (1973)	CBA/H-T6	47	6.0
6–7 day	Takagi and Sasaki (1976)	A/He	1406	2.8
8–11 day	Ford and Evans (1973)	F_1	1924	0.6
10–11 day	Yamamoto et al. (1973a)	CF	149	0.0
10–11 day	Speed and Chandley (1981)	Q	571	0.7
10–14 day	Fabricant and Schneider (1978)	pooled data on five different strains. (see Table 5.4.)	149	0.0
10–14 day	Ford and Evans (1973)	F_1	239	0.0

* *In vivo* fertilization.
† *In vitro* fertilization.

TABLE 5.4. *Incidence of aneuploidy in first cleavage and early- to mid-gestation embryos of the mouse in relation to maternal age*

Reference	Strain	Maternal age and % aneuploidy		
Tease (1982)	3H1	2–4 month (0.38%)	11–13 month (3.08%)	
Maudlin and Fraser (1978b)	TO	2–4 month (3.3%)	8–10 month (7.5%)	
Yamamoto et al. (1973b)	CF	3–5 month (0.0%)	11–16 month (2.5%)*	
Speed and Chandley (1981)	Q	1½–2 month (0.7%)	9–12 month (1.2%)	
Gosden (1973)	CBA/H-T6	1 month (6.0%)	8–12 month (21.0%)	
Fabricant and Schneider (1978)	CBA	2–5 month (0.0%)	7–10 month (13.6%)	11 month + (0.0%)
	NZB/J	2–5 month (0.0%)	7–10 month (0.0%)	11 month + (not tested)
	C57B1/6J	2–5 month (0.0%)	7–10 month (0.0%)	11 month + (5.5%)
	A/J	2–5 month (0.0%)	7–10 month (9.0%)	11 month + (0.0%)
	C3H/H3J	2–5 month (0.0%)	7–10 month (5.6%)	11 month + (8.0%)
Max (1977)	CBA	2–3 month (0.0%)	5–6 month (0.0%)	11 month + (7.4%)

* Figure recalculated by Gosden and Walters (1974).

conceived by young and aged mothers have been made, the data show age-related increases, but only for some strains (Table 5.4). The data are variable, as were those reported earlier for aging oocytes analysed at metaphase II. For the CF and Q strains respectively Yamamoto *et al.* (1973*a,b*) and Speed and Chandley (1981) reported modest increases over the ages tested when F_1 fetuses were karyotyped. The TO mice, used by Maudlin and Fraser (1978*b*) showed more than a doubling in aneuploidy in first-cleavage embryos with age while Tease (1982) found an almost tenfold rise for the 3HI strain. The embryonic difference in both studies was found to be due solely to differences between the female complements of the two groups. Interestingly, analysis of the female data of Maudlin and Fraser (1978*b*) showed that there was no significant difference between groups in the incidence of monosomy; the difference lay only in the incidence of trisomy. From the studies of Gosden (1973), Max (1977), and Fabricant and Schneider (1978), there is an indication of a maternal age effect also for the CBA strain. Aneuploidy levels in the latter study appeared to reach a peak in females between 7 and 10 months, but then declined in animals aged 11 months and more. Indeed, these studies, which gave information for a total of five different mouse strains, showed similar responses in four of them, and although a modest further increase was found in the remaining strain over 11 months, this was certainly not of the magnitude of the exponential increase observed with human maternal aging. Indeed, from all the studies on germ cell and early cleavage embryos in the mouse and other species, it would appear that no animal investigated has yet been found to show spontaneous levels of aneuploidy approaching those of man, and a maternal-age effect which is quite so strong and continuous. Table 5.5 gives the available data concerning aneuploidy levels in embryos from animal species other than the mouse, and it can be seen that in all of them

TABLE 5.5.   *Incidence of aneuploidy in early embryos of domestic and laboratory species other than the mouse (Aneuploidy frequency obtained by doubling the hyperploid counts)*

Species	Embryonic stage	Reference	No. embryos analysed	% aneu-ploidy
Chinese hamster	4–8 cell	Binkert and Schmid (1977)	226	0.9
Sheep	2–8 cell	Long and Williams (1980)	89	9.0
Rabbit	5½ day	Fechheimer and Beatty (1974)	463	1.7
Pig	10 day	McFeeley (1967)	88	0.0
Rat	11 day	Butcher and Fugo (1967)	410	0.0
Cow	12–16 day	McFeeley and Rajakoski (1968)	12	0.0

(with perhaps the exception of the sheep) numerical errors are as rare as they are in the mouse.

If we accept 2 per cent as the figure for aneuploidy at conception in the mouse, the comparable figure for man of perhaps 10 per cent as suggested by Ford (1975) and as indicated by the human sperm analysis data of Martin *et al.* (1982), is of an order of magnitude greater. Moreover, the strong exponential rise in aneuploidy encountered in older women does not appear to be shown by old mice. Rather, the indications in the mouse from several different studies, both on oocytes, F_1 embryos and liveborn progeny, are for a maternal-age increase in aneuploidy followed perhaps by a plateauing at very advanced ages. Data are limited, however, and further comparative studies at young and very advanced maternal ages are required.

Can the mouse therefore justifiably continue to be used as a model system for human aneuploidy and for the study of the age effect? Several authors believe that it can. Schneider and Kram (1981) have argued that the mouse is still the best species available for testing because it is extremely well characterized genetically, and has several inbred strains whose reproductive capabilities have been well documented. Moreover some strains are available whose reproductive capabilities are altered. For example, the CBA strain has a significantly shorter reproductive life-span than other inbred mouse strains. Furthermore, fetal aneuploidy can be found with increasing frequency in many strains, as a function of maternal age in the mouse, even though a levelling-off after a certain time may occur. Epstein (1981) believes that because of cost, rapidity of breeding, availability of genetic markers, the existence of interesting aneuploid phenotypes, and availability for producing aneuploid progeny in relatively high yields (Gropp and Epstein 1982), the mouse has become, and will remain, the animal of choice for the development of a model system for aneuploidy.

The usefulness of the mouse as a model in tests to determine the effects of X-irradiation, chemicals, and other agents in inducing extra nondisjunction or aneuploidy will be illustrated in the next two chapters.

6. The induction of aneuploidy by physical agents

RADIATION

Evidence for induction

In 1972, the United Nations Scientific Committee on the Effects of Atomic Radiation (UNSCEAR) recommended that future research in the field of radiation genetics should include studies on the mechanisms of nondisjunction and its induction by irradiation of germ cells in experimental organisms.

Ionizing radiation has, of course, been known for many decades to cause gene mutation and chromosome breakage: the need for research into its effects in terms of aneuploid induction has developed more recently, and for two main reasons. One is the knowledge, accumulated over the past twenty years or so, of the remarkably high incidence of chromosome anomalies, particularly aneuploids, among human conceptuses and live-borns. The other is the ever-increasing exposure of human beings, both occupationally and environmentally, to noxious agents, including ionizing radiation, having the potential to increase even further the levels of aneuploidy within the population.

In this section, we will consider the experimental evidence that exposure of cells to ionizing radiation can result in an increase in the frequency of aneuploidy.

The discovery that ionizing radiation is trisomigenic is a very old one. The increase in aneuploidy frequency following the X-ray treatment of *Drosophila* females was reported some years before Müller's historic discovery of the mutagenic action of X-rays in 1927. At least in *Drosophila*, X-rays remain one of the most potent trisomigens yet discovered. First we will review the evidence for this, and assess the extent to which radiation is effective as a trisomigen in other organisms. Later we will consider the possible ways in which such increases might be brought about.

Induction in non-mammalian species

Although the effect of X-rays on mutation has been extensively studied in lower eukaryotes, especially in relation to the repair of radiation-induced damage, little work has been carried out on aneuploid induction. This is largely because there are few meiotic systems suitable for the study of aneuploidy and, of these, only two appear to have been used to study the effect of ionizing radiation. Parry *et al.* (1979*b*) have shown that very high

TABLE 6.1.   *The effect of X-rays on the frequency of sex chromosome aneuploids in Drosophila.*

	Control				Irradiated				Reference
	Regular progeny		Exceptional progeny		Regular progeny		Exceptional progeny		
	♀♀	♂♂	XXY ♀♀	XO ♂♂	♀♀	♂♂	XXY ♀♀	XO ♂♂	
	3606	3734	n.d.*	1	1452	1431	n.d.*	24	Mavor (1924)
	5264	5499	4	7	1547	1462	5	34	
	7711	7528	1	5	1771	1558	8	43	Mavor (1922)
	1743	1726	0	1	512	467	2	12	
	501	545	n.d.	0	77	71	n.d.	10	
	68 353†	64 124	15	61	3577	3289	10	64	Anderson (1924)
	25 021†	22 524	17	70	18 414	16 828	73	397	Demerec and Farrow (1930*a*)
	20 871	16 317	2	23	34 849	30 960	17	338	Demerec and Farrow (1930*b*)
	13 580	10 136	1	14	26 917	24 071	17	266	

* Not detectable.
† Control data of Safir (1920) used by Anderson (1924).

doses of X-rays can induce aneuploidy at meiosis in yeast, and initiated studies into the effect of X-rays in relation to DNA synthesis prior to meiosis. Yeast cells were irradiated either immediately after plating on sporulation medium (which induces meiosis) or twelve hours after plating. The latter had completed DNA synthesis before being exposed to radiation. Exposure to X-rays increased aneuploidy in both populations of cells, but there was greater induction at low doses in cells which had had a longer exposure to sporulation medium before treatment. Griffiths (1979) reported increased aneuploidy following treatment with γ-rays in *Neurospora*. In view of the uncertainty which exists over the mechanism of radiation-induced aneuploidy (see below) it would be valuable to extend these studies, both by continuing the work in yeast, and by extending it to other microbial systems.

The trisomigenic effect of X-rays has been studied much more extensively in insects, especially *Drosophila*. Aneuploid induction was first reported by Mohr (1919) in a locust, *Decticus verrucivorus*, and by Mavor (1921, 1922, 1924) who studied induction in *Drosophila*. The data obtained by the first *Drosophila* workers in this field are summarized in Table 6.1. It can be seen that X-irradiated *Drosophila* females exhibit an obvious and statistically significant increase in progeny with sex-chromosome aneuploidy, the frequency of both exceptional XO males and XXY females being increased compared to controls.

Subsequent work has been facilitated by the fact that the time course of gametogenesis in *Drosophila* has been established by tritium autoradiography timing studies, both in the male (Chandley and Bateman 1962) and in the female (Chandley 1966), so that known meiotic stages can be exposed to irradiation and their sensitivities to aneuploid induction assessed. In the *Drosophila* female, King and his fellow workers have investigated oogenesis in detail (King *et al.* 1956; King 1957; Koch *et al.* 1967) and classified the developing oocyte into fourteen morphologically distinct stages. Dävring and Sunner (1973, 1976, 1977) have studied meiosis cytologically and their results, combined with the timings established by Chandley (1966), have been put together in a schematic way in Table 6.2.

Each ovary of a female *Drosophila* consists of about twelve egg tubes or ovarioles. Each ovariole is divided into an anterior germarium and a posterior vitellarium and contains oocytes at various stages of development. In a newly emerged female there are no oocytes which have developed beyond stage 7. If such a female is irradiated, the first eggs laid by her will have been exposed to radiation whilst they were immature, stage 7 oocytes. A mature female, on the other hand, contains a mature, stage 14 oocyte in each ovariole with a stage 7 oocyte immediately behind it in the vitellarium. If, therefore, a mature three- or four-day-old female is irradiated, then the first thirty or so eggs laid will correspond to oocytes irradiated whilst they were at stage 14.

TABLE 6.2. *A summary of oogenesis in* Drosophila *based on King* et al. *(1956), Dävring and Sunner (1975, 1977), and Chandley (1966)*

Development stage	Morphogenetic description	Cell division stage	Development time
Stem line oogonia	Cell division gives one stem cell and one cystoblast	Mitosis	
Cystoblast	Cell division gives 16 interconnected cystocytes from one cystoblast	4 mitotic divisions	5 days (King) 3 days (Chandley)
Cystocytes	The 16 cystocytes give 1 oocyte and 15 nurse cells		
Stage 1 oocyte	Cell cluster incompletely surrounded by follicle cells	Leptotene	
Stage 2 oocyte	First egg chamber formed when follicle cells surround germarial cyst	Leptotene/ zygotene	3 days
Stage 3 oocyte	Chromosomes more deeply stained	Pachytene	
Stage 7 oocyte	Egg chamber has enlarged nurse cells = oocyte in size	Diplotene	
Stage 14 oocyte	Fully grown primary oocyte with complete appendages	Modified diplotene (King) / Metaphase I (Dävring and Sunner)	12 hours or less
Egg in egg tube Newly deposited egg		Anaphase I Metaphase II	

In the early experiments, summarized in Table 6.1, the flies were normally irradiated within 24 hours of emergence, so that the effect of X-rays on immature (up to stage 7) oocytes was being studied. Patterson *et al.* (1932) and Uchida (1962) studied the effect of X-rays on virgin females which had been aged for various periods before irradiation. In these experiments the aging process alone had no effect on the frequency of aneuploidy, but irradiation produced a significant increase in the frequency of sex chromosome exceptions. The extent of the induction increased with increasing age of the females. This interaction of radiation and aging is represented graphically in Fig. 6.1 from which it can be seen that the older the stage 14 oocyte irradiated, i.e. the longer it had been stored in the female, the more susceptible it was to radiation-induced aneuploidy.

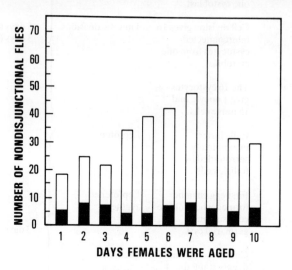

Fig. 6.1. The effect of aging *Drosophila* oocytes before irradiation on the production of aneuploids. Solid bars represent controls, open bars represent treated series. (From Patterson *et al.* (1932).)

The effect of an increasing X-ray dose on sex chromosome aneuploid frequency has been studied by several workers (Mavor 1924; Demerec and Farrow 1930*b*; Traut 1964, 1970; Parker and Busby 1973). Figure 6.2 presents the more extensive results of Traut (1964) and Parker and Busby (1973). In the latter experiments only immature oocytes were irradiated whilst, in the former, all stages were irradiated. As might be expected, increased exposure resulted in increased aneuploidy. The shape of the dose response curve is relevant to considerations of the mode of action of X-rays, and we shall return to consider the kinetics of induction in a later section.

The trisomigenic effect of X-rays in *Drosophila* is not confined, however, either to the sex chromosomes or to the female sex. Bateman (1968) studied aneuploid induction for the second chromosome following X-ray treatment. Oocytes which are aneuploid for either major autosome normally give rise to inviable zygotes when they are fertilized, but by using male isochromosome stocks, Bateman was able to 'rescue' aneuploid oocytes because, in such stocks, the sperm is itself often aneuploid. Thus fertilization of an aneuploid oocyte by a complementary aneuploid sperm restores a balanced genotype and gives a viable zygote (see Fig. 6.3). Clark and Sobels (1973) and Savontaus (1977) also detected X-ray induced aneuploidy of the major autosomes using this technique.

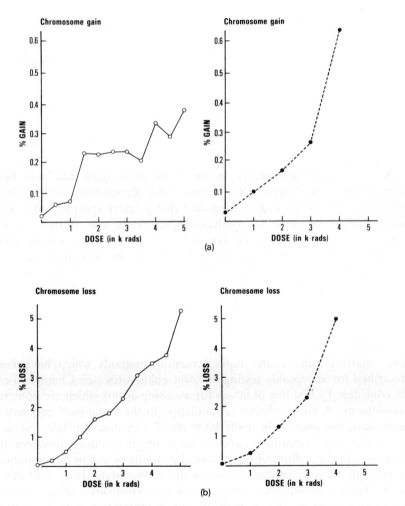

Fig. 6.2. The kinetics of X-ray induced aneuploidy in *Drosophila*; (a) chromosome gain, (b) chromosome loss. Left-hand graphs—data of Traut (1964); right-hand graphs—data of Parker and Busby (1973).

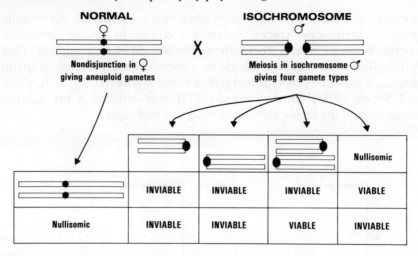

Fig. 6.3.   Illustration of the isochromosome method for detecting aneuploidy, for either of the major autosomes, arising in female *Drosophila*.

X-ray induced aneuploidy in the *Drosophila* male has also been demonstrated (Sävhagen 1960; Strangio 1961; Zimmering and Wu 1963). The data of Strangio (1961) suggested that primary spermatocytes were particularly susceptible to irradiation, but doubt was thrown on this conclusion by Zimmering and Wu (1963). They were able to distinguish nondisjunction of the X-Y bivalent from X-ray mediated exchange between the sex chromosomes, and concluded that the X-ray effect on primary spermatocytes could largely be accounted for by the induction of rearrangements between the X and Y chromosomes.

Induction in mammals

The relatively simple and rapid screening methods which have been described for aneuploidy testing in lower eukaryotes (see Chapter 3) can be considered a first line of attack for assessing agents which are potential trisomigens. A more reliable extrapolation to the aneuploidy problem in man, can, however, only really be made if a mammalian test system is used. The main limitations to aneuploidy testing in mammalian systems lie, however, in the tedium of the technical procedures and in the amount of time required to accumulate a worthwhile body of data. In spite of this, a number of X-ray studies have been carried out, principally in the mouse, and principally on germ cells. A small amount of information does also exist on the effects of irradiation and other exogenous agents on chromosome segregation in somatic cells.

THE RESPONSE OF GERM CELLS IN THE MALE

Little is known of the possible consequences, in terms of aneuploidy induction, of exposure of the human testis to ionizing radiation. Suggestive evidence has come from data of Boué *et al.* (1975) in France, of a significantly greater occupational exposure to radiation among the fathers of chromosomally abnormal fetuses, and particularly those with numerical anomalies. In a study carried out by Smithers *et al.* (1973) in London, however, no chromosomally abnormal children were produced by men who had previously been therapeutically irradiated to the para-aortic and iliac nodes for the treatment of testicular tumours. Altogether, a total of 52 normal children were produced by 34 fathers between the ages of 25 and 45 who had received an estimated testicular dose of 50–70 rad during exposure of the nodes to 4000 rad given over a three to five week period. Most of the children were conceived from spermatozoa which were stem cell spermatogonia at the time of X-ray exposure.

In the mouse, much of the data concerning the effects of ionizing radiation on the male germ line has been contributed by L. B. Russell and her co-workers of the Oak Ridge National Laboratory, Oak Ridge, Tennessee, USA. The relative sensitivities of all stages of spermatogenesis from spermatogonia to spermatozoa have been tested using the system described in Chapter 5, for the detection of viable sex chromosome anomalies among liveborn progeny (for review, see Russell 1976). The data concern both pre-meiotic and post-meiotic irradiations. The latter obviously cannot produce aneuploidy by true nondisjunction but loss of a paternal sex chromosome resulting in the recovery of X^MO animals (carrying a maternal X chromosome) was found by Russell and Saylors (1962, 1963) to be readily induced by the treatment of spermatids or spermatozoa. The peak frequency was found in early spermatids, these yielding approximately three times as many X^MO cases as did irradiated spermatozoa and about eight times as many as mature spermatocytes. Because the results parallelled in several ways those for dominant lethals and translocations, it was considered likely that most of the induced paternal sex chromosome losses resulted from breakage events.

For pre-meiotic irradiation, 'germinal cell selection' becomes an important factor in determining the results of the experiments and the level of X-ray exposure chosen. From the early work of Oakberg (1960), it is known that while spermatozoa and spermatids are relatively insensitive to cell killing by irradiation, primary spermatocytes will be killed in high numbers by doses in the hundreds of rads (Oakberg and di Minno 1960). For spermatogonia, the population is mixed with regard to killing: extreme sensitivity is shown by Type B cells but some Type A cells are very radio-resistant. Spermatogonial irradiation, in fact, was found by Russell and Montgomery (1974) not to yield aneuploid offspring, even when an X-ray dose of 600 rad was used. For primary spermatocytes, however,

Russell and Saylors (1963) recovered a total of 16 exceptional animals out of 5630 offspring classified (0.53 per cent) after 200 rad X-rays, compared with only one out of 3059 (0.06 per cent) in controls. The difference was significant. Of the 17 exceptions, one (in the treated group) was an XXY male which could have arisen by true nondisjunction. All the others were X^MO exceptions which had arisen by paternal sex chromosome loss. Russell (1976) believed that failure to recover such X^MO exceptions following the irradiation of spermatogonia might have been because breakage in these cells did not lead to recoverable single chromosome loss, either because no sister chromatid joining occurred or, if it did, because this led to cell division failure. Alternatively, she argued, sex chromosome loss might have occurred, but the resulting XO and OY cell lines so produced might have been inviable in the testis. That monosomics might act as spermatogonial cell lethals in the mouse has also been suggested by Searle (1975). In experiments designed to induce nondisjunction or chromosome loss in spermatogonia by irradiation, aneuploids have not been detected among the metaphase I spermatocytes derived from the treated cells (Ford *et al.* 1969). Szemere and Chandley (1975) also failed to recover aneuploid metaphase II spermatocytes in male mice irradiated with 100 or 200 rad to spermatogonia. Significant increases in numbers of aneuploid second metaphases were, however, reported when preleptotene or pachytene primary spermatocytes were treated at the 200 rad dose level. More recently, Speed and Chandley (1981) reported a threefold overall increase in chromosome abnormalities among F_1 fetuses derived from irradiated spermatogonia compared with controls, but the increase in aneuploids alone did not reach a level of significance. This apparent failure to recover aneuploids among the descendant germ cells or F_1 offspring derived from treated spermatogonia is interesting in the light of early reports by Griffen and Bunker (1964, 1967) of rare primary trisomics among the F_1 progeny of X-irradiated male mice treated in the sperma-togonial stages. Doses of 350 and 700 rad were used in their experiments. Lyon and Meredith (1966) suggested, however, that the extra element in the karyotype could in fact have been a translocation product, the 41 chromosome mice in question being tertiary and not primary trisomics. The reports were made prior to the advent of banding methods for the identification of chromosomes. Griffen and Bunker (1964, 1967) were themselves convinced, from the meiotic configurations, that they were in fact observing trivalents and were thus dealing with primary trisomics.

The most recent attempt to induce sex chromosome aneuploidy in males by irradiation, appears to have been made by Danford (1982). Using X-ray doses of 50 and 100 rad to pre-meiotic cells (sampled as spermatozoa six weeks after treatment) she produced two suspected aneuploid offspring out of 588. A total of 1103 progeny from control samples and breeding stocks showed no aneuploid progeny. The test-system involved the use of a strain of mouse carrying the X-linked Tabby marker. XO or XX females were used to detect, among the F_1 progeny, sex chromosome aneuploidy arising at the first meiotic division in the male parent.

One of the few other mammalian species in which males have tested for aneuploidy induction using ionizing radiation is the northern vole, *Microtus oeconomus* and the results are somewhat intriguing. Tates (1979) and Tates *et al.* (1979) reported significant increases in X-ray induced sex chromosome nondisjunction in early experiments, but were unable later to repeat their findings (Tates and de Vogel 1981). They suggested that genetic changes might have occurred in their laboratory stocks since the first tested animals were collected from the wild. Selection against bad breeders or nondisjunction-prone individuals were suggested as possible explanations for subsequent failures to detect nondisjunction.

THE RESPONSE OF GERM CELLS IN THE FEMALE

From the early work of Russell (1968, 1976) it is known that irradiation of the mouse oocyte during the dictyate stage in the adult, or during the preleptotene to diplotene stages in the fetus and neonate, can result in the induction of OX^P individuals. Only in the latter experiments, using a dose of 200 rad, was the arrangement of markers such as also to detect the possible occurrence of nondisjunction leading to $X^M X^M Y$: none was observed. With regard to this loss of the maternal X chromosome, the predictyate prophase stages taken together were less sensitive than the dictyate stage in mature follicles. When irradiation of mouse oocytes was carried out between dictyate and diakinesis (Reichert *et al.* 1975), hyperploidy was observed at metaphase II at doses up to 200 rad, but the increases were not significant.

In recent years, investigations into the effects of X-rays on mouse oocytes have been more concerned with the problem of whether aneuploidy is induced at very low dose levels. In particular, attention has been paid to the question of whether aged oocytes have a susceptibility to low dose ionizing radiation which is relatively greater than that for young oocytes.

The concern has arisen out of human epidemiological studies concerning the radiation induction of trisomy in women who have received low dose diagnostic or therapeutic X-rays during their reproductive years. Uchida (1979) has reviewed sixteen epidemiological studies on the association between maternal preconception irradiation exposure and subsequent birth of a child with an aneuploid genotype. The results, however, are conflicting. Six studies showed a significant positive association; seven a non-significant positive association and three a non-significant negative association. One of the negative bodies of data came from the large series of children examined for Down syndrome in Hiroshima and Nagasaki (Schull and Neel 1962). A subsequent cytogenetic study of these children

(Awa 1975) found no autosomal trisomy in 2885 children of exposed parents and 1090 controls, and a non-significant increase (eight out of 2885 versus one in 1090) in the frequency of cases with supernumerary sex chromosomes.

Because some of the human data indicate a higher incidence of Down syndrome among children of irradiated older mothers (Alberman *et al*. 1972), recent experimental studies in the mouse have attempted to determine whether or not there is an age effect on the X-ray induction of nondisjunction, but again, the data are equivocal.

One of the earliest studies was that made by Yamamoto *et al*. (1973*b*). These authors presented data which, they claimed, demonstrated the greater sensitivity of aged mouse oocytes compared to young oocytes for a 5-rad X-ray exposure. The authors emphasized the further increase in aneuploidy seen in old treated mice compared with old controls.

On re-examination of the data, and the way in which they had been interpreted, Gosden and Walters (1974), however, could find no justification for such conclusions. They criticized the statistical procedures employed, the grouping of mosaics with true aneuploids and the inclusion of control data which had been obtained in earlier experiments (Yamamoto *et al*. 1973*a*) when environmental conditions could well have been very different. In the view of Gosden and Walters (1974), the differential effect of low-dose X-irradiation on young and aged mouse oocytes remained unproven.

Subsequent studies, using F_1 fetal karyotyping for analysis, have given results consistent with this view (Max 1977; Strausmanis *et al*. 1978; Speed and Chandley 1981). None of these studies has recorded any increases in aneuploidy either for young or aged mothers exposed to irradiation between doses of 2 and 16 rad. Tease (1982), scoring one-cell embryos, was also unable to detect a greater sensitivity among aged oocytes compared to young but, in this case, was able to demonstrate induction following irradiation. Both young and old oocytes gave the same linear increases in aneuploidy and structural chromosome damage when treated, at the preovulatory stage (diakinesis/metaphase I), with doses of 10 to 50 rad. No increases were found, however, when dictyate oocytes were irradiated 11 days prior to culture (Tease 1981). It thus appears that oocyte stage at the time of irradiation might be the critical factor in determining sensitivity to X-ray induced nondisjunction.

The only study which would seem to provide evidence for a differential X-ray sensitivity of mouse oocytes with age remains that of Uchida and her colleagues. They found an increase in aneuploidy at metaphase II following the treatment of young dictyate oocytes, with doses of 10 to 30 rad (Uchida and Lee 1974), with an even greater increase for aged oocytes (Uchida and Freeman 1977). The increases found, however, were not linear with dose.

ZYGOTE IRRADIATIONS

In addition to studies in which germ cells have been irradiated, some experiments have been aimed at inducing aneuploidy by irradiation of zygotes. Russell and Saylors (1962, 1963) for example, showed that irradiation at various times between sperm penetration of the egg and the first cleavage, can yield characteristic frequencies of sex chromosome loss (see Russell 1976, for review). Using the appropriate sex-linked markers, instances of both maternal and paternal sex chromosome loss were detected although the vast majority of exceptions were X^MOs. The peak incidence of X^MO and OX^P that resulted from treatment during this interval were in fact many times higher than those observed after irradiation of any of the germ-cell stages in either sex. Indeed the frequency of paternal sex chromosome loss in the early pronuclear stage was about ten times greater than that induced by irradiation of spermatozoa in the vas deferens of epididymis. The XO frequency was highest after irradiation given early on the day of fertilization (when X^POs as well as X^MOs were produced) and somewhat lower when the animals were exposed later in the day when only X^MOs were recovered (Russell and Saylors 1963). Irradiation during subsequent cleavages (1½–4½ days after fertilization) were ineffective in increasing the frequency of XO individuals. The change of sensitivity, it was suggested (Russell 1976), may have been related to DNA synthesis, which starts somewhere between the early and late pronuclear stages: the high XO frequencies may have occurred prior to initiation of the S-phase. It is not known, however, to what extent breakage events were responsible for such losses.

HUMAN BLOOD LYMPHOCYTE IRRADIATIONS

Finally, mention should be made of studies in man which have used the blood lymphocyte to evaluate the effect of low doses of radiation on the segregation of human somatic chromosomes. It is of course not possible to obtain experimental data on aneuploidy induction with X-rays in human germ cells because of obvious practical and technical difficulties, and the results on somatic cells cannot necessarily be extrapolated to meiotic segregation. Nevertheless, the data are of some interest.

The main study on irradiated lymphocytes has been carried out by Uchida *et al.* (1975) using a low dose of 50 rad X-rays, chosen in order to minimize the level of chromosome breaks. Seven fathers and seven mothers of 21-trisomic children and equal numbers of age and sex-matched controls with no known history of Down syndrome among close relatives were chosen for study.

The frequency of hyperploid cells in the irradiated samples was four times greater than in controls (0.1 per cent compared to 0.025 per cent). Furthermore, significant increases in the frequency of aneuploidy were noted in unirradiated lymphocytes exposed to irradiated cell-free plasma or serum. The X and No. 21 chromosomes appeared particularly susceptible to nondisjunction and the authors suggested that this might be relevant to the frequent occurrence of those two chromosomes in aneuploidy, particularly in mosaics, in the general population and among abortuses. X-ray induced aneuploidy, chiefly chromosome loss, has also been reported for somatic cells of the mouse (Wennström 1971).

The mode of action of radiation in aneuploidy induction

When the parts of the cell division machinery which are affected by radiation and which consequently give rise to aneuploidy through malfunction are considered, it is found that most of the information comes from work on *Drosophila*, where investigations have been very extensive. This account inevitably concentrates therefore on that organism. Evidence from other sources is often very sketchy, but it is included here in an attempt to see whether the conclusions reached from the *Drosophila* work are equally valid for other systems. The inclusion of this evidence also has the advantage that it highlights those areas where more data are needed,

TABLE 6.3.　*Summary of hypotheses to account for the trisomigenic action of radiation in* Drosophila

Proposed effect of radiation	Consequence	Reference
Reduction in chromosome pairing	Increased nonconjunction	Savontaus (1975)
Reduced crossing-over in distal regions	Partial desynapsis	Chandley (1968)
Increased crossing-over in proximal regions	Increased difficulty of separating homologues	Müller (1954)
Damage to centromere	Not specified	Clark and Sobels (1973)
On second division of meiosis	Increased nondisjunction at second division	Anderson (1931) Bateman (1968)
Induction of chromosome interchanges	Recovery of rearranged chromosomes and correlated aneuploidy of non-homologous chromosomes	Parker (1969)
On spindle fibres	Nondisjunction following damage to microtubules	Traut (1970)
Increased chromosome stickiness	Difficulty in separation of homologues	Rapoport (1938)

because, despite a research effort extending over many decades, the mechanism of action of radiation in inducing aneuploidy is still not fully understood.

The fact that most of the data come from *Drosophila* experiments means, however, that caution must be exercised in applying the conclusions, in a general way, to all organisms. It is quite possible that *Drosophila*, which has several unusual meiotic features, may be a special case. For example, the chromosomes at meiosis in the oocyte are held together in a chromocenter (Dävring and Sunner 1973, 1976) which might predispose the chromosomes to undergo nondisjunction in a manner peculiar only to the *Drosophila* female. The regular disjunction of non-homologous chromosomes, which can occur with a high frequency under some experimental conditions (p. 50), might typify this special behaviour.

There have been various suggestions made to account for the action of X-rays in inducing aneuploidy and these are summarized in Table 6.3. Some of the hypotheses are general and unspecific, and no obvious experiments to test them have yet been devised. Others are quite specific and have been tested experimentally. Conflicting claims have been made over the mode of action of X-rays and no one hypothesis is able to account for aneuploid induction in all organisms, nor need it do so, because the various hypotheses are not mutually exclusive. It might be expected that X-rays will act to disrupt cell division in more than one way.

The effect of radiation on crossing-over and chromosome pairing
IN *DROSOPHILA*
The fact that X-rays can affect the frequency of crossing-over in *Drosophila* has been recognized from the early days. Mavor (1923) observed a reduced frequency of crossing-over in the X-chromosome following irradiation and suggested that the increase in sex-chromosome aneuploidy might be related to this in some way. Mavor considered that the effect was indirect, because X-rays produced a reduction in crossing-over throughout a six-day period which was considerably longer than the time considered necessary for crossing-over to proceed to completion.

The effect of X-rays is not always to reduce crossing-over. Mavor reported that, in the same experiments in which crossing-over in the X chromosome was reduced, crossing-over in the second chromosome increased. Müller (1925), using multiply-marked chromosomes, subsequently showed that irradiation increased crossing-over in regions adjacent to the centromere but decreased it in more distal regions. Chandley (1968) has since confirmed this differential effect of X-rays on crossing-over in different regions of the X chromosome and shown that the distal reductions extend to the median segments at higher doses.

Anderson (1931) studied the frequency with which non-crossover chromosomes were involved in the formation of aneuploid zygotes. He observed that crossover chromosomes in sex-chromosome aneuploids were found at only 60 per cent of the control level. These studies were extended by Bateman (1968) and Savontaus (1975) who studied the effect of X-rays on aneuploidy and crossing-over by irradiating different meiotic stages in the female. Both authors showed that irradiation of oogonia resulted in a marked reduction in subsequent meiotic crossing-over and that the chromosomes in aneuploids derived from irradiated oogonia were very largely non-crossovers. Irradiation of oocytes in meiotic prophase resulted in the greatest induction of aneuploidy and these aneuploids too were often non-crossovers but to a lesser extent than those arising from irradiated oogonia (Savontaus 1975).

Two alternative explanations have been advanced to account for the correlated effects of irradiation on crossing-over and aneuploidy. In the first, the increase in aneuploidy is seen as a result of radiation-induced interference with chromosome pairing; the reduction in crossing-over is then a consequence of reduced pairing. Bateman (1968) suggested that disomics arising from irradiated oogonia were produced as a result of asynapsis. Chandley (1968) proposed that X-rays interfered with chromosome pairing in distal regions and as X-ray doses increased so pairing inhibition spread towards the median segments. Savontaus (1975) suggested a causal relationship between increased aneuploidy and decreased crossing-over. The different proportions of crossover chromosomes arising from the irradiation of different meiotic stages suggested to her that more than one mechanism might be involved, but she also suggested that the primary effect of irradiation in inducing aneuploidy was to bring about a failure of chromosome pairing.

The second explanation was proposed by Müller (1954) who took into account the differential effect of X-rays on crossing-over in different regions. He attributed increased nondisjunction to difficulties of chromosome separation following increased crossing-over in the centromere regions.

IN OTHER ORGANISMS

An indication of the effect of radiation on chromosome pairing in other organisms has been obtained indirectly from experiments in which levels of recombination or chiasma formation have been monitored. There have been several experiments of this kind. Chiasma formation, in *Lilium* and *Tradescantia*, appears to be sensitive to irradiation during two short stages of meiosis (Lawrence 1961*a,b*). Irradiation at pre-leptotene in both these species resulted in a reduced frequency of chiasmata per bivalent, while irradiation in early pachytene caused increases. A similar differential effect of X-rays on chiasma frequency in *Schistocerca gregaria* was found by Westerman (1967).

In *Chlamydomonas*, Lawrence (1965) detected similar changes in recombination frequency using low doses of radiation, but at high doses, recombination was reduced at all stages.

These results are consistent with the idea that irradiation just before meiosis (pre-leptotene) interferes with chromosome pairing, and consequently reduces crossing-over, but later experiments in which radiation was applied at both these sensitive stages throws doubt on this hypothesis (Lawrence 1968). When both pre-leptotene and zygotene/pachytene were irradiated, there was an increase in recombination, as though zygotene alone had been irradiated. If irradiation at pre-leptotene interfered with pairing, this result would not be expected, since pairing failure should preclude stimulation of recombination at a later stage.

A more direct approach which has been tried is to measure the effect of irradiation on the frequency of univalents seen at metaphase I in cytological preparations, since, if irradiation does interfere with pairing during prophase these might be expected to increase in number. Wennström (1971), Walker (1977) and Chandley (unpublished) have all estimated the frequency of autosomal univalents and of unattached X-Y

TABLE 6.4. *Frequency of univalents and unattached X and Y chromosomes following irradiation of male mice*

Stage irradiated	Dose (rad)	No. of Cells analysed	No. (+%) of cells with autosomal univalents	No. (+%) unattached X–Y	Reference
L. Pachytene/	50	300	12 (4.0)	16 (5.3)	
diplotene	100	300	13 (4.3)	20 (6.7)	Wennström
	200	300	32 (10.7)	26 (8.7)	(1971)
	400	300	46 (15.3)	22 (7.3)	
Control	—	1150	43 (3.7)	91 (7.9)	
Spermatogonia		185	4 (2.2)	6 (3.2)	
Preleptotene		246	10 (4.1)	11 (4.5)	
Leptotene		500	15 (3.0)	19 (3.8)	
Zygotene	200	500	20 (4.0)	30 (6.0)	Walker
Pachytene		500	26 (5.2)	33 (6.7)	(1977)
Diplotene		300	25 (8.3)	15 (5.0)	
Diakinesis		83	3 (2.4)	2 (2.4)	
Control	—	600	10 (16.7)	14 (23.3)	
Preleptotene		200	8 (4.0)	4 (2.0)	
Leptotene		200	12 (6.0)	6 (3.0)	
Zygotene		200	5 (2.5)	5 (2.5)	
Pachytene	200	200	7 (3.5)	6 (3.0)	Chandley
Diplotene		200	14 (7.0)	2 (1.0)	(unpublished)
Diakinesis		200	12 (6.0)	6 (3.0)	
Control	—	200	8 (4.0)	8 (4.0)	

chromosomes at metaphase I following X-irradiation of pre-meiotic cells from male mice. The data are summarized in Table 6.4. Both Wennström and Walker reported a significant increase in the frequency of autosomal univalents after irradiation at pachytene, but found no significant effect when radiation was given at earlier stages. Chandley, however, was unable to detect a significant effect at any stage. The frequency with which the X and Y chromosomes were present as univalents also gave conflicting results. Walker (1977) detected a significant increase following irradiation of pachytene and diplotene but neither Wennström (1971) nor Chandley (unpublished) were able to detect any such increases.

The results are therefore inconclusive. However, even if the frequency of univalents at meiosis were increased by radiation, it does not follow that this would necessarily result in increased aneuploidy in the gametes. It will be recalled (p. 71) that Polani and Jagiello (1976) and Speed (1977) were unable to detect increased aneuploidy at metaphase II in aged oocytes of the mouse, in spite of recording increased univalent frequencies at metaphase I. This indicates either that oocytes carrying univalents are lethal, or that the univalents disjoin correctly. Failure of pairing in spermatocytes of the mouse (Beechey 1973; Purnell 1973) and man (Chandley *et al.* 1976*a*) has already been suggested as a cause of germ-cell arrest in males and this may also hold true in oocytes. If so, it would provide a most effective filter mechanism by which potential aneuploid products could be selected against in the female germ line.

The effect of radiation on centromere regions
IN *DROSOPHILA*

Since there is good evidence for involvement of the centromere in chromosome movement, it would not be surprising if one important effect of radiation were to be on this region. Müller (1954) observed that the inducing effect of X-rays was not correlated with the size of the chromosomes, and this suggested to him that the target for the X-rays was not the chromosome itself. The effect of X-rays in inducing crossing-over in the centromere regions of *Drosophila* led him to propose that the target for X-rays was indeed the centromere region. Grell *et al.* (1966) also studied the relationship between radiation-induced aneuploidy and chromosome size and confirmed that the frequency of chromosome gain was independent of size. It will be recalled that crossing-over in the centromere region was implicated in the processes leading to chromosome loss in some meiotic mutants (p. 47) so it might be thought that the frequency of chromosome loss, too, would be independent of chromosome size. However, Grell *et al.* (1966) found that X-ray induced chromosome loss was correlated with chromosome length. They suggested that induced chromosome loss arose from chromosome breakage followed by fusion to form dicentric bridges. The X-ray induced chromosome loss in spermatids and spermatozoa of male mice has also been suggested by Russell (1976) to involve breakage events resulting in sister chromatid rejoining.

IN OTHER ORGANISMS

The effect of non-ionizing irradiation on chromosome movement has been studied directly, by several groups of workers, in experiments in which a microbeam of ultraviolet light was applied to particular parts of the cell during division. These experiments will be considered more fully in the section dealing with the effects of non-ionizing irradiation on the spindle apparatus (p. 115), but consideration will be given here to those experiments in which the kinetochore region of a chromosome was selectively irradiated.

There is a great deal of evidence that the kinetochore is a structural element of considerable importance in chromosome movement (see reviews by Luykx 1970 and Bajer and Molè-Bajer 1972). There is also direct evidence that irradiation of a chromosome in this region can influence chromosome movement.

Izutsu (1961*b*) studied the effect of ultraviolet light on grasshopper spermatocytes during metaphase I and anaphase I. When one kinetochore in a bivalent was irradiated during metaphase I, the irradiated bivalent moved towards the pole to which the unirradiated kinetochore was pointing. The treated bivalent did not regain its position on the metaphase plate, but, in most cases, disjunction of the homologues at anaphase I was not affected. Irradiation of the kinetochore during early anaphase I did not affect chromosome movement.

Bajer (1972) studied the effect of microbeam and microslit irradiation on mitotically dividing endosperm from *Haemanthus katherinae* and showed that whilst irradiation of the kinetochore at early anaphase had no effect on chromosome movement, microslit irradiation of mid-anaphase caused it to slow down or stop.

To what extent these disturbances are temporary and reparable, and whether or not they can lead to aneuploidy, will be considered further when the effects of radiation on the spindle fibres are considered.

The effect of radiation on second-division nondisjunction

IN *DROSOPHILA*

Several authors have claimed that X-rays can produce an increase in nondisjunction at the second division; normally this is very infrequent in spontaneously occurring aneuploids in *Drosophila* (p. 37). It will be recalled that nondisjunction at the second division is identified in genetic analysis by homozygosity for proximally located genetic markers (p. 36) and Anderson (1931) observed a relatively high frequency of homozygosity for the X-linked mutations 'forked' and 'garnet' after X-rays. This led him to suggest that second-division nondisjunction of the X chromosome had occurred with an increased frequency. However, Parker *et al.* (1974) threw considerable doubt on Anderson's conclusion that X-rays could induce second division errors. In experiments using genetic markers more closely linked to the centromere than forked and garnet, they were unable to find any evidence for increased second-division errors following X-ray treatment.

Bateman (1968) has also claimed increased second-division nondisjunction of the autosomes, although in this case the results were also somewhat ambiguous. The centromere markers in one chromosome arm were homozygous as expected for a second-division error, but the centromere markers in the other were heterozygous. Bateman therefore suggested that nondisjunction at the second division was, in some way, dependent on crossing-over near the centromere. In cases where proximal crossing-over is thought to have occurred it is not possible to conclude whether the segregational error occurred at the first or second division.

IN OTHER ORGANISMS
No experiments appear to have been carried out in other organisms to see whether or not radiation treatment increases the frequency of nondisjunction at the second meiotic division.

The effect of radiation in promoting chromosome interchanges
IN *DROSOPHILA*
Parker and his co-workers have proposed and extensively tested the hypothesis that X-rays induce aneuploidy by inducing chromatid interchanges between non-homologous chromosomes. Following an interchange, the segregation of the chromosomes can be affected, resulting in the simultaneous gain and loss of pairs of chromosomes. This hypothesis was developed from a study of X-ray induced detachment of attached-X chromosomes, and we will first consider these experiments.

Parker (1969) observed that irradiation of attached-X stocks resulted in their detachment, and this event was associated with chromatid interchange with the acrocentric fourth chromosome. The hypothesis was advanced that, when immature oocytes were irradiated, an interchange resulted in the formation of a 'quasi-bivalent'. At the following anaphase the involved chromosomes segregated away from each other, i.e. there was 'directed disjunction' of the involved elements, so that they moved to opposite poles. The non-involved fourth chromosome segregated independently of the quasi-bivalent.

The 'directed disjunction' hypothesis has several expected consequences, some of which are illustrated in Fig. 6.4. Firstly, the directed disjunction of the quasi-bivalent will result in the recovery of one or other half-translocation in an oocyte, but not both. As expected, following irradiation of immature, stage 7, oocytes, the reciprocal products of the interchange are not found in the same meiotic product. Secondly, after an interchange between the attached-X and a fourth chromosome, aneuploid gametes which are nullo-X diplo-4, and others which are attached-X nullo-4, are expected. That is, if aneuploid gametes are formed as a result of interchanges between chromatids of non-homologous chromosomes, then coincident aneuploidy of the X and fourth is to be anticipated. Traut and Scheid (1971) and Parker and Busby (1973) noted that about 20 per cent of nullo-X gametes were diplo-4 and about 75 per cent of diplo-4 gametes were nullo-X.

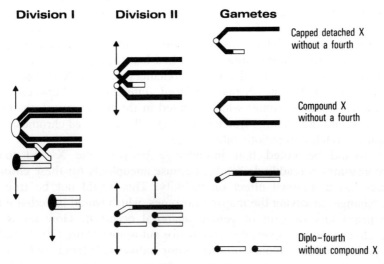

Fig. 6.4. Some of the consequences of interchange between a compound-X and a fourth chromosome in *Drosophila* following X-ray irradiation of immature (stage-7) oocytes. This diagram illustrates how half-translocations (but not reciprocal translocations) and co-incident aneuploidy for the X and fourth chromosomes are generated. Not all segregational possibilities are included in this diagram. (After Parker and Williamson (1974).)

This model has been extensively analysed and tested (see Parker and Williamson 1974, 1976), and it has emerged that the directed disjunction of the involved elements only seems to occur when immature oocytes are irradiated. Following irradiation of mature, stage 14, oocytes, both involved elements can segregate to the same pole resulting in the recovery of reciprocal translocations. Busby (1971) has explained this finding by suggesting that the segregation of the chromosomes is already determined in stage 14 oocytes and interchanges between chromatids do not disturb this.

This model can successfully explain how X-ray treatment of attached-X stocks induces meiotic aneuploidy, but what is the evidence to support the idea that the model can also explain aneuploid induction in karyotypically normal *Drosophila* females? If interchanges occur between free X chromosomes and the fourth chromosome, the situation is more complicated because the two Xs may have already formed a crossover bivalent; any interchange will therefore result in the formation of a quasi-trivalent or quasi-multivalent (Parker and Busby 1973). Since about 95 per cent of X bivalents have one or more crossovers, this will apply to the majority of interchange chromosomes. Parker and Busby (1973) studied X-ray induced nondisjunction of free X chromosomes in crosses which allowed the detection and recovery of several different, fertile aneuploid types. They found that 22 per cent of XO males were trisomic for chromosome 4 and contained two maternal fourths. They also recovered a very high frequency of altered fourth chromosomes which had lost their genetic markers, as though they had been involved in interchanges with the X chromosome. They concluded that X-rays induced interchanges between non-homologous chromosomes which resulted in the formation of aneuploid gametes, and that 'a large fraction, perhaps all, of *induced* chromosome-4 trisomy is related to meiotic interchange'.

It should be noted that interchange between the X and fourth chromosomes is readily detectable because aneuploidy for these chromosomes has a minimal effect on viability. This would not be true for interchanges involving the major autosomes, which would be lethal if any significant loss or gain of genetic material occurred. How far is the interchange model applicable to the major autosomes? Puro (1978) studied X-ray induced aneuploidy for the major autosomes. Irradiated females were mated to specially constructed males so that any oocytes which had interchanges between the second and third chromosomes would give rise to viable zygotes if they were fertilized by sperm of a complementary genotype. Puro was able to demonstrate that oocytes with a rearrangement between the second and third chromosomes did occur but the directed disjunction of the X and fourth, characteristically observed when immature oocytes were irradiated, did not occur in the case of autosomal interchange. Savontaus (1977) demonstrated that, after X-ray treatment, 41 per cent of offspring derived from eggs containing two second chromosomes were simultaneously aneuploid for the X chromosome, a result which can be explained by assuming that an interchange between the X and the second chromosome is a principle cause of the aneuploidy (see Fig. 6.5).

Despite this demonstration that interchanges do occur, it is not clear to what extent the induction of aneuploidy for the major autosomes in karyotypically normal stocks can be accounted for by chromosome interchange. Since induction is observed in such crosses in which any large

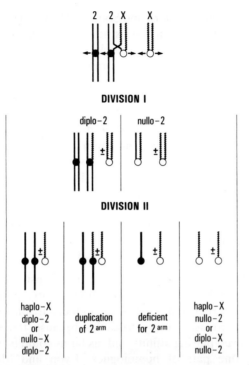

Fig. 6.5. Consequences of the second chromosome nondisjunction caused by an interchange between an X and a second chromosome, assuming the 'directed disjunction' of the elements involved in the interchange to opposite poles. (From Savontaus (1977), with permission.)

rearrangements would certainly be lethal, it is clear that induced aneuploids must either possess very small rearrangements (so that viability is not affected) or be induced by meiotic errors not involving chromatid interchanges.

The kinetics of X-ray induced chromosome loss would seem to support the hypothesis that it arises from chromosome breakage. There is general agreement that the dose response curve for chromosome loss shows a greater than linear response with increasing dose, which is compatible with the idea that the induction of loss requires two hits. There is less unanimity over the shape of the dose response curve where chromosome gain is concerned. In particular, there has been some disagreement over the response at low doses. Traut (1970) claimed that immature oocytes showed a threshold, X-ray doses below 1000 R having no detectable effect. This has been interpreted by Traut (1971) as evidence that X-rays increase aneuploidy by affecting the spindle apparatus. Parker and Busby (1973) threw some doubt on this interpretation by demonstrating a statistically significant effect of 1000 R. Clark and Sobels (1973) also could not find any evidence for a threshold when they studied the induction of autosomal aneuploidy using isochromosome stocks. They showed a significant increase in autosomal aneuploidy for X-ray doses as low as 250 R. Sobels (1979) demonstrated induction following 125 R but the dose–response curve was complex; there was a plateau for doses up to 1000 R with a linear increase for doses ranging from 1000 R to 3000 R.

Clark and Sobels (1973) obtained a different dose response when different stages were irradiated. Induction after irradiation of stage 7 oocytes followed single hit kinetics whereas irradiation of earlier stages gave results that followed two hit kinetics. These observations led Sobels (1979) to conclude that several mechanisms were involved in the induction of disomic gametes. It should be noted, however, that these experiments were carried out on isochromosome stocks in which chromosome pairing is not necessarily normal.

IN OTHER ORGANISMS

There is virtually no evidence either for or against the idea that the interchange hypothesis can be extended to other organisms. Griffen and Bunker (1964, 1967) reported seven X-ray induced trisomics in the mouse, which they regarded as primary trisomics because the supernumerary chromosome showed pairing affinity and, as far as could be seen, complete homology with one pair of homologues. Lyon and Meredith (1966) described tertiary trisomics following irradiation, and suggested that the first three aneuploids reported by Griffen and Bunker might also have been tertiary trisomics, i.e. they contained a rearranged chromosome as the 'extra' one.

If analysis of chromosome homology in trisomics revealed that the additional chromosome was frequently a chromosome rearrangement, then serious consideration would have to be given to the idea that X-ray treatment induces aneuploidy by promoting chromosome interchanges. Unfortunately, at the present time, there are too few cases described to enable a conclusion to be drawn.

The effect of radiation on the spindle apparatus

IN *DROSOPHILA*

The hypothesis that X-rays affect the division spindle in *Drosophila* is based on very indirect evidence. There are two main arguments. Firstly, Traut (1970) claimed the existence of a threshold below 1000 R, and a threshold might be expected if X-rays had an inhibitory effect on spindle formation either by affecting the spindle itself or by inhibiting the processes leading up to its formation. Secondly, experiments in which Traut (1971) studied the effect of splitting the X-ray dose and of extending the exposure time by lowering the dose rate, showed that both fractionation and protraction of the dose led to a lowering in the incidence of chromosome loss. This might be expected if, for example, there was some restitution of damage over the extended time scale involved in dose fractionation and protraction. The frequency with which gametes gained a chromosome, on the other hand, was only affected by lowering the dose rate; splitting the dose into two equal parts, separated by an interval of as much as five hours, had no effect. The lack of response to dose fractionation is a feature unique to the induction of nondisjunction and suggested to Traut that the action of X-rays was a 'physiological' one, possibly on the spindle.

There are some difficulties with this idea. If Parker is correct in his conclusion that X-rays do not affect disjunction at the second division (p. 110), then it is difficult to reconcile the observation that irradiation can influence events at anaphase I, several days after treatment, while producing no effect at anaphase II, which follows only a few hours later (Parker and Busby 1973).

IN OTHER ORGANISMS

There is better evidence that radiation affects the spindle from other experimental systems in which the effects of non-ionizing radiation are observed directly. The problems of focusing X-rays so that only selected parts of a cell are irradiated do not appear to have been overcome, but apparatus has been developed which allows microbeams of ultraviolet light to be generated (Uretz *et al.* 1954). Many groups have studied the consequences, for chromosome movement, of irradiating the spindle apparatus (see, for example, Uretz and Zirkle 1955; Izutsu 1961*a,b*; Inoué 1965; Forer 1965, 1966, 1969; Bajer 1972; Bajer and Molè-Bajer 1972).

When the spindle fibres are irradiated with a microbeam, a discrete area of reduced birefringence is produced, and this is thought to be the result of changes in the microtubules of the spindle (Bajer 1972), although Forer (1965, 1966, 1969) has argued that the microtubules are not responsible for the birefringence.

Irradiation of a mitotic cell during the stage when the spindle is being formed results in inhibition of spindle formation (Uretz and Zirkle 1955). Irradiation of a cell at later stages was studied by Izutsu (1961*b*) who irradiated one side of the spindle during metaphase. When the microbeam was applied to the spindle near the kinetochore, the bivalent moved towards the pole on the unirradiated side. The effect was transitory and the bivalent regained its position on the metaphase plate. Izutsu interpreted this as showing that a bivalent on the metaphase plate is under traction from the spindle fibres and that the damaged spindle was quickly repaired. This observation that irradiated spindles at metaphase give rise to chromosome movement has, however, been challenged (Forer 1966 and see discussion by Luykx 1970).

When the spindle is irradiated during anaphase, chromosome movement can be slowed down or stopped (Izutsu 1961*b*; Forer 1966; Bajer 1972). In these cases, the laggard chromosome is stopped only temporarily, and soon catches up. When the irradiation is sufficient to slow down chromosome movement it can be shown that the irradiated region of the spindle has a reduced number of microtubules (Bajer 1972), and it seems reasonable to conclude that one effect of radiation is to disrupt the spindle fibres thereby affecting chromosome movement.

Although, then, there is clear evidence that ultraviolet light (UV) does affect chromosome movement, the evidence from both microbeam and micromanipulation experiments (Nicklas and Staehly 1967; Nicklas 1967, 1969) points to an ability of the spindle to repair induced damage over a short time interval. The effect of ionizing radiation on aneuploid production, in contrast, lasts several days and, therefore, it seems unlikely that a large part of aneuploid induction can be attributed to effects on the spindle fibres.

It should be noted, however, that Bajer (1972) has recorded and photographed an interesting case of UV-induced nondisjunction by first irradiating a bivalent to induce chromosome stickiness, and then irradiating the spindle fibres near the kinetochore to affect chromosome movement. Under these circumstances the irradiated chromosome was seen to undergo nondisjunction.

Irradiation of the spindle can have consequences for other parts of it beside the microtubules. The components of the cell responsible for determining the number of spindle poles can be affected by radiation (Luykx 1970). There are several reports that radiation can induce a multipolar condition within mitotically dividing cells (Henshaw 1940; Levis and Marin 1963; Fetner and Porter 1965; Rondanelli *et al.* 1967) and we have seen (p. 45) that multipolar spindles can give rise to aneuploidy with simultaneous and coincident loss of several chromosomes (multiform aneuploidy) (Böök 1945).

The effect of radiation on the chromosome surface

IN *DROSOPHILA*

Another 'physiological' consequence of radiation is to increase the 'stickiness' of chromosomes, and it has been suggested (Rapoport 1938) that this can account for increased aneuploidy following X-rays. Rapoport was led to make this suggestion because he observed that aneuploidy for the X and fourth chromosomes of *Drosophila* were not independent, the correlation suggesting to him that a change in both chromosomes had occurred simultaneously.

IN OTHER ORGANISMS

There is no doubt that chromosome stickiness can be induced by a wide range of experimental treatments, including irradiation (Lea 1955; Bacq and Alexander 1961). For example, Izutsu (1961*b*) irradiated selected parts of a chromosome with a UV microbeam and induced stickiness. There is, however, little evidence that irradiation of a chromosome alone can lead to aneuploidy. In cases of induced stickiness, the chromosomes may disjoin, albeit reluctantly, or, in cases where the chromosomes are very sticky, a restitution nucleus may be formed as cell division fails (White 1937). However, as mentioned above, Bajer (1972) did observe nondisjunction of an irradiated chromosome when the spindle fibres were also irradiated. The UV dose necessary to induce stickiness was 3–4 times less than that necessary to affect chromosome movement by spindle irradiation.

Stickiness is a descriptive term applied to cases where chromosomes are held together at anaphase by means other than exchange of genetic material. McGill *et al.* (1977) examined sticky chromosomes using an electron microscope, and suggested a possible material basis for stickiness. They found strands interconnecting the chromosomes which they suggested came about from entanglement of chromatin fibres during chromosome condensation. If this is the explanation for stickiness, then it might be expected that entanglement of fibres would be greater for larger chromosomes and therefore stickiness should be correlated with chromosome size. If this is so, the observation that nondisjunction frequency occurs independently of chromosome size argues against extensive induction through increased stickiness.

Radiation-induced aneuploidy : an overview

What conclusions can be drawn, about radiation-induced aneuploidy, from this array of experimental evidence? The evidence that ionizing radiation is a trisomigen in *Drosophila* is overwhelming, but is much less convincing in mammals, where most of the data come from experiments on the mouse. These latter experiments are obviously more relevant in assessing the trisomigenic danger of radiation to man.

In terms of long-term genetic hazard in man, the germ cells most at risk are the spermatogonial stem cells in males and the resting diplotene oocytes in females. For aneuploidy induction by irradiation, the data would indicate that the risk of inducing trisomy is very small; there is no clear evidence of induction after the irradiation of mouse spermatogonia and, in mouse dictyate oocytes, although small increases in hyperploidy have been recorded by some authors, others have been unable to demonstrate any increase either in young or old females. The preovulatory (diakinesis/metaphase I) stage of oocyte development does appear, however, to be more sensitive than the dictyate stage to X-ray induced nondisjunction in the mouse. The bulk of the experimental data from the female mouse would appear, moreover, not to support the idea that the aged dictyate oocyte is any more susceptible to radiation-induced aneuploidy than the young.

X-monosomy can, however, be readily induced by irradiation in the mouse at almost all stages of germ-cell development in both males and females. The early spermatids appear to be particularly sensitive; the spermatogonia, however, are insensitive. Moreover, there would appear to to exist in the mouse, a relatively high probability of loss from the zygote of the paternally contributed sex chromosome, some time between fertilization and the first cleavage. Maternal sex chromosome loss can also be induced by irradiation at this time but this is a much rarer event.

What are the underlying causes of radiation-induced aneuploidy? Most of the data aimed at answering this question come from *Drosophila* and generalizations should only be made with caution. The germ cells of *Drosophila* are more resistant to killing by radiation than those of mammals, and this could be why levels of induction in *Drosophila* are so much greater than in mammals. If, as Parker suggests, most aneuploids in *Drosophila* result from chromosome interchange, the proportion from this source in mammals may be much smaller simply because the lower doses do not induce sufficient breaks while, at higher doses, the cells die.

Whilst bearing this cautionary note in mind, the following conclusions seem justified:

1. There is some evidence that radiation can affect chromosome pairing, at least when some meiotic stages are irradiated, and that this may lead to aneuploidy. In mammals, extensive pairing failure can lead to sterility but it is not clear to what extent, if at all, this is a barrier to aneuploid production through occasional pairing failure.

2. At least in *Drosophila*, centromere regions are of some importance in that radiation causes increases in crossing-over in these regions, and aneuploids inherit chromosomes which are recombinant near the centromere more often than expected.

3. There is doubt as to the effect of radiation on nondisjunction at the second meiotic division, but here the evidence is confined to *Drosophila*. It is not clear how generally any conclusion applies to other species.

4. In *Drosophila*, the most rigorously tested hypothesis is Parker's idea that induced aneuploidy arises as a consequence of chromosome re-arrangements. It has stood up to test in a convincing way, but two doubts remain. First, it is not clear whether it can account for all X-ray induced aneuploidy in *Drosophila*. Secondly, it is not clear whether this mechanism is peculiar to *Drosophila*, especially considering that the chromosomes in the oocyte are held together in a chromocentre which might facilitate rearrangement between non-homologous chromosomes.

5. Although it has been directly demonstrated that radiation affects the spindle and chromosome stickiness, it is not clear if these changes result in increased nondisjunction. The ability of the spindle to repair itself argues

against the idea that any long-lasting effect of radiation is attributable to its direct action on the spindle.

6. It is believed that most of the induced paternal sex chromosome losses in mammalian germ cells result from breakage events. To what extent breakage events are responsible for such losses in the mouse zygote is not however known.

This review will have highlighted the fact that very little is known for certain about the underlying cause or causes of aneuploid induction following exposure to ionizing radiation. Investigation of many unanswered questions is still necessary. One problem arises from the fact that most investigations using genetic methods have been carried out in *Drosophila*, and this may not be a representative organism. It is important to extend the analysis to other organisms which should be carefully selected so that at least some of the alternatives can be distinguished. For example, genetic methods using well marked chromosomes can simultaneously monitor the effect on crossing-over (and therefore infer the importance of desynapsis), and the effect on nondisjunction at the second division. Experiments using an organism like *Neurospora crassa*, would also enable the importance of chromosome rearrangements to be assessed because these are easily detected in this species.

The susceptibility of cells in the mammalian germ line to killing by radiation means that radiation-induced aneuploidy may be less important in man than aneuploidy induced by other trisomigens. The issue is important enough, however, to justify further experiments to estimate the likely extent of aneuploid induction by radiation in mammals, and by extrapolation, in man.

TEMPERATURE VARIATION

Compared with the extensive literature concerning aneuploidy induction by ionizing radiation, that dealing with induction by heat or cold treatment is limited.

Early experiments in *Drosophila*, carried out by Hildreth and Ulrichs (1969) showed that variations in the frequency of X-chromosome nondisjunction occurred when females were subjected to various temperatures during prolonged periods of aging preceding egg deposition. During treatment, females were placed on minimal medium which reduced oviposition. Newly eclosed virgin females were exposed to low temperatures (10 °C) and increases in the frequency of X-chromosome aneuploidy occurred in females subjected to 10 °C compared to those at 25 °C. However, since females exposed to 10 °C showed reduced oviposition compared to those at 25 °C, Hildreth and Ulrichs (1969) were unable to distinguish between an aging effect on the eggs and an effect of temperature. This is important, because it has been shown that suppressed oviposition itself results in increased sex chromosome aneuploidy (Traut and Schröder 1978; Traut 1980).

More extensive data were later obtained by Tokunaga (1970*a,b*). The parental females were newly eclosed virgins which were 'pre-treated' by being placed on minimal food at 25 °C for three days, a procedure used by Hildreth and Ulrichs (1969). After this common pre-treatment period, the females were treated by being placed, either at 10 °C or at 25 °C, on minimal food for zero days (control series) or three days, or one, two, or three weeks (experimental series). After the respective temperature and aging treatment, each female was placed on normal food and mated at 25 °C for a 13-day brood period.

Each of the four experimental series at 25 °C showed a trend towards a higher frequency of nondisjunctional exceptions with increasing aging: in all groups, the frequency of exceptions was higher in the later than in the earlier broods compared with controls. A similar pattern of effect was found when females were aged at the lower temperature of 10 °C, but in contrast to the 25 °C series, for which there was no clear-cut difference between first-day treated and control broods, all 10 °C treated series produced first-day broods with very greatly increased frequencies of exceptional flies. Interestingly, the sex ratio of nondisjunctional exceptions (XXY:XO), induced by low temperature, tended to be 1:1, suggesting that the exceptions did not include many cases of chromosome loss. This is in sharp contrast to the effects of X-rays on aneuploidy induction in *Drosophila*, where an excess of XO males over XXY females is characteristically found. Spontaneous aneuploidy in *Drosophila* is also characterized by a great excess of XO males to XXY females.

Another difference between X-ray and low temperature induced nondisjunction in *Drosophila* appears to be that the low temperature effect is restricted to the retained mature eggs while the X-ray effect is found to be very high also in later broods. The eggs involved in producing the first-day broods, in the experimental series, it was suggested, corresponded to vertebrate eggs in a state of post-ovulatory overripeness. The effects of aging, apparent in later broods, were based on influences exerted on younger egg stages.

In subsequent studies, Tokunaga (1970*b*) showed that the low temperature-induced nondisjunction of the X chromosomes in retained mature eggs occurred exclusively at the first meiotic division. Also, in addition to X-chromosome exceptions, other exceptions occurred as a consequence of nondisjunction of the autosomes. This positive correlation between X-chromosomal and autosomal aneuploidy is not, however, specific for the low temperature effect. It has also been observed in spontaneous nondisjunction (Hall 1970) and with some meiotic mutants causing nondisjunction (Sandler *et al.* 1968).

The low temperature treatments of Tokunaga (1970*b*) also caused failure of disjunction for whole chromosome sets, giving rise to polyploidy. It was not determined, however, whether the high susceptibility of retained mature eggs to low temperature induction of nondisjunction was due to failure of cell division, e.g. a spindle defect or to a defect in the chromosomes, e.g. the kinetochores.

In another series of cold treatment experiments in *Drosophila*, carried out by Leigh (1979), a very high frequency of sex chromosome and second chromosome aneuploidy was also recorded in eggs laid in the first brood after storage of females for seven days at 8–10 °C. A much lower frequency was found among the progeny of eggs laid on the 2nd–4th days after cold treatment. As in Tokunaga's experiments, similar frequencies of cold-induced gains and losses were recorded.

The effect of cold treatment on the *Drosophila* male has also been studied (Tokunaga 1971). Males were exposed to 10 °C prior to mating following which the seventh and eighth day broods showed highly significant increases in sex chromosome aneuploidy. In this case there was an excess of XO males compared to XXY females. This result indicated the existence of a cold-sensitive period during spermatogenesis. If the timing of spermatogenesis established by Chandley and Bateman (1962) was unaltered by cold temperature treatment, then the primary spermatocytes would appear to be the sensitive stage to cold treatment.

Tokunaga (1971) points out several differences between the effect of low temperature on males and females. Perhaps the most striking of these are the absence of any effect on autosomal nondisjunction in the male and the equality of XXY and XO exceptions in the female.

Nondisjunction has also been found to occur with a high frequency at elevated temperatures. When Grell (1971*b*) treated developing *Drosophila* females for 24 hours at 35 °C (treatment to oogonia), a primary nondisjunction rate for the X chromosomes approximately twice as great as that obtained with 4000 rad of X-rays was found. No increase in autosomal nondisjunction was found, however. Grell (1971*b*) noted that there was a positive correlation between X-nondisjunction frequency and an increase in noncrossover X tetrads, suggesting that such an increase was a necessary prerequisite for heat-induced nondisjunction. Heat, it had been found, failed to increase autosomal noncrossovers.

The effect of elevated temperature on nondisjunction in the X-bivalent appeared to be an indirect one, since a period of about six days elapsed between treatment (at the late oogonial stage) and effect (at the first meiotic division).

In mammals, few studies into the effects of temperature on nondisjunction or chromosome loss appear to have been made. Seasonal fluctuations in human aneuploidy, possibly attributable to temperature variations have, however, been reported (e.g. Nielsen and Friedrich 1969).

One interesting experimental study was carried out by Karp and Smith

(1975). They subjected cultured mouse oocytes to a nine-hour incubation at 37 °C, by which time most eggs were at metaphase I. They then transferred them to a temperature of 23 °C where they remained for the next 12 hours. Within one hour of transfer, there was loss of the precise orientation of the chromosomes on the spindle. After the full 12 hours at 23 °C, the oocytes were reincubated at 37 °C when they regained their proper orientation and proceeded (in the majority of cases) to metaphase II. A total of 108 control (cultured throughout at 37 °C) and 125 experimental oocytes were analysed at metaphase II. Among the latter group, 9.6 per cent were found to be hyperploid, with chromosome counts ranging from 21 to 28, giving an aneuploidy level of 19.2 per cent. Hypoploidy was also found, but excluded, as many such oocytes were thought to have arisen by artefactual chromosome loss. In addition, a significant number of cells with diploid or quasi-diploid complements were found, suggesting an effect on suppression of the first polar body, similar to that recorded in the low temperature experiments in *Drosophila* described earlier (Tokunaga 1970*b*).

An extensive search of the literature has revealed no other studies into the effects of heat or cold shock on chromosome segregation in mammalian germ cells. One brief report by Nebel and Hackett (1961) does, however, suggest that elevated temperature can cause interference with the pairing of homologues at meiotic prophase in the mouse and reduce the total number of synaptonemal complexes observed in primary spermatocytes. Such pairing irregularities might in turn, lead perhaps to some nondisjunctional events.

For cultured mammalian mitotic cells, there are data showing that a lowering of temperature can influence chromatid separation. Hela cells grown at 29 °C instead of the usual 37 °C show a gradual increase in mitotic index and an increasing frequency of anomalous division accompanied by mitotic chromatid nondisjunction (Rao and Engelberg 1966). Moreover, Nagasawa and Dewey (1972) have reported an increase in polyploidy in Chinese hamster cells subjected to cold temperature *in vitro*. These responses to cooler temperature appear identical to those elicited by low doses of colcemid and certain oestrogens (Rao and Engelberg 1967) (see Chapter 7).

The mechanisms of induction of cold- or heat-induced nondisjunction may or may not be the same. For elevated temperature, there are indications from the *Drosophila* studies of Grell (1971*b*) and from the studies of Nebel and Hackett (1961) in the mouse, that the primary effect of heat may be in causing desynapsis in the prophase bivalents. It is known from other studies, that heat can bring about reductions in recombination in *Drosophila* (Plough 1917; Stern 1926; Chandley 1968) and in chiasma frequencies in other insect species (Henderson 1962; Church and Wimber 1969). It has been suggested that heat-induced reductions in recombination

(Chandley 1968) and chiasma frequency (Church and Wimber 1969) arise also through effects on synapsis or on the synaptonemal complex. The heat-sensitive stages would appear to be pre-leptotene to zygotene–early pachytene.

Buss and Henderson (1971*a,b*) reported experiments on *Locusta migratoria* which are also consistent with the idea that elevated temperature affects chromosome pairing. Following exposure of leptotene or zygotene to heat, interlocked bivalents were found. In most cells these interlocked bivalents showed unipolar orientation of each bivalent. In contrast with unipolar orientations of normal bivalents, the unipolar orientation of the interlocked bivalent was stable and usually resulted in double nondisjunction at anaphase I. Comparable effects of cold treatment have been found in *Melanoplus* (Henderson *et al.* 1970).

Böök (1945) studied the effect of cold shock on *Triton*. Many cytological disturbances were reported including areas of the testes where synapsis of the chromosomes had failed or was not maintained, and multipolar spindles which led to multiform aneuploidy (p. 45).

Alternatively, heat treatment may have effects on the spindle. A high frequency of tetraploid cells was in fact observed by Westra and Dewey (1971) when heat was applied to Chinese hamster mitotic cells in culture. Nagasawa and Dewey (1972) suggested that the polymerization processes involved in spindle formation were probably inhibited at high temperatures as they are with colchicine, and perhaps also with cold. Cold treatment has been shown to destroy the mitotic spindle in frog cells (Hertwig 1898) and Inoué (1952) found that spindle-fibre birefringence was abolished below 5 °C in *Chaeopteris* eggs. Roth (1967), also, has demonstrated that the mitotic apparatus microtubules of giant amoebae disassemble completely at 2 °C. Electron microscope studies on mammalian cells, however, indicate that certain microtubules of the mitotic apparatus are more resistant to cold disruption than are others. Using fibroblasts of rat kangaroo, Brinkley and Cartwright (1975) showed that after 30 minutes in an ice-bath, interpolar tubules were preferentially disrupted at metaphase and early anaphase, while chromosomal tubules were considerably more resistant. Conversely, interpolar microtubules present at late telophase were resistant to disruption by cold. The findings suggest that the lability of a single class of microtubule may change during the course of mitosis.

Conclusions

It is likely therefore that temperature variation (either heat or cold shock) can induce aneuploidy, at least in lower organisms where the effects have been studied the most. Like ionizing radiation, it is likely that there is more than one underlying cause of this induction. It is likely that prolonged exposure to either high or low temperature affects both chromosome pairing and chromosome segregation through effects on the spindle. However, the effects of temperature and ionizing radiation appear to differ in one important respect. At least some of the trisomigenic effect of radiation can be attributed to chromosomal breakage leading to chromosome loss. The observed excess of XO over XXY progeny in radiation experiments on both *Drosophila* and the mouse suggest breakage followed by X chromosome loss as the mechanism responsible for this phenomenon. At least in the female, on the other hand, temperature-induced aneuploidy appears to arise by an effect on chromosome segregation, in that monosomics and trisomics are found with equal frequency.

Another difference between the two types of treatment in the *Drosophila* female, would appear to be time and duration of sensitivity. The period of marked sensitivity, following high or low temperature shock, is much more limited than that seen following radiation. Whether this difference is characteristic of the two types of treatment or is merely a reflection of dose (low X-ray dose giving a temporary effect like that found with heat, high X-ray dose giving a more long-lasting effect) is not known.

7. The induction of aneuploidy by chemical agents

We have seen that ionizing radiation can increase aneuploidy and, at least in *Drosophila*, probably does so by acting directly on the chromosomes and causing breakage. Aneuploidy is thus seen to be a secondary consequence of rearrangement.

Radiomimetic chemicals may also act in this way although the use of numerical anomalies as endpoints in chemical mutagenesis has been limited. One worrying possibility arises when consideration is given to the induction of aneuploidy by chemical agents, including those which occur as pollutants of the environment. As with ionizing radiation, some may act directly on the chromosomes. Others, however, may give negative results in mutation tests because they do not act directly on DNA. In separate aneuploid-induction tests, however, they might prove to be potent trisomigens by virtue of the fact that they can affect the proteins involved in cell division. By studying agents which are known, or thought to interact with cell division proteins, the possible existence of such compounds could be tested for and verified. Then, a start could be made in evaluating the efficiency of methods used in their detection.

A search of the literature reveals that chemical agents of many different kinds have been tested for their response in terms of aneuploidy induction. The data concern a variety of species and are scattered widely in the literature in a large number of publications. In order, therefore, to review the subject in some kind of logical and coherent way, and to reduce the task to manageable proportions, we have decided to classify the chemical compounds according to their probable modes of action. Three principal categories of compound wil be dealt with:

Those which act directly on chromosomes or DNA.

Those which act on tubulin and microtubules.

Those which act on the spindle in other ways.

COMPOUNDS ACTING DIRECTLY ON CHROMOSOMES OR DNA

Alkylating agents

Several alkylating agents have been tested in lower eukaryotes and found to be trisomigenic. Morpurgo *et al.* (1979) tested methyl methane sulphonate (MMS) on quiescent and germinating conidia of *Aspergillus nidulans*. Conidia in both states gave large increases in the frequency of mitotic nondisjunction. Earlier Shanfield and Käfer (1971) had tested

N-methyl-N-nitro-N-nitrosoguanidine on diploid conidia of *Aspergillus* and reported increases in $2n + 1$ conidia resulting from the treatment. The numbers were small, however, and the authors concluded that larger samples were needed before any firm conclusions could be drawn. Recently Sora *et al.* (1982) have described a system in yeast which distinguishes aneuploidy arising at the first and second divisions of meiosis. In addition diploid meiotic products resulting from complete failure of the second division can be identified. MMS treatment resulted in induction of both aneuploidy and diploidy but, in these particular experiments, it was not possible to determine which meiotic division gave rise to the aneuploidy.

Several workers have studied the effects of alkylating agents in *Drosophila*. Petrova (1976) and Sobels (1979) reported experiments in which neither chromosome loss nor chromosome gain was induced following ethyl methanesulphonate (EMS) treatment. In Sobels' case the effectiveness of the exposure to EMS was monitored by examining the frequency of sex-linked lethal induction in flies exposed in a similar manner. On the other hand Mittler (1976) and Foureman (1979) reported increased chromosome loss but no increase in chromosome gain following EMS treatment.

In the mouse, EMS and MMS have been reported to give increases in aneuploidy at metaphase II in spermatocytes treated at the pre-leptotene stage (Szemere 1978). Following an intraperitoneal injection of EMS, seven out of 414 metaphase II cells were hyperploid (1.7 per cent) and nine out of 395 (2.3 per cent) were hyperploid following MMS treatment. These findings compared with no aneuploid cells in 400 examined from control animals.

Significant increases in X^M loss following treatment of dictyate oocytes of the mouse have been reported by Generoso *et al.* (1973) using IMS (isopropylmethanesulphonate). No induction of trisomy was, however, found in this study.

Cattanach (1964), in an early study into the effects of the alkylating agent triethylenemelamine (TEM) on germ cells of the mouse, recovered, among the F_1 sons being tested for translocations, one sterile male which had arisen following treatment to a primary spermatocyte in the sire. The animal, on karyotyping, turned out to have 41 chromosomes, an aneuploid thought to have arisen by nondisjunction at one of the meiotic divisions in the treated father. The extra chromosome was interpreted as a No. 16, but since the study was made before chromosome banding techniques were developed, this remains unconfirmed. Trisomy 16 in the mouse is now known, however, to be one of the few autosomal trisomics which are compatible with adult life (Gropp and Epstein 1982).

In a further study with TEM, Cattanach (1967b) demonstrated a clear increase in paternal sex chromosome loss following treatment to spermatids and spermatozoa. The late spermatids were particularly sensitive. As the post-meiotic stages are also the sensitive ones in the male mouse to TEM-induced dominant lethals (Bateman 1960) and translocations (Cattanach 1957), it is possible that breakage events underly the induced chromosome losses. Also in mice, Schleiermacher (1968) treated males with triethylene iminobenzoquinone (Trenimon) and cyclophosphamide (Cytoxan). Cytological examination revealed an increased frequency of univalents at metaphase I, especially after pre-leptotene treatments but no corresponding frequency of aneuploid cells at metaphase II. Röhrborn and Hansmann (1971) examined metaphase II oocytes after treating mice with Trenimon and Cytoxan and reported extremely high frequencies of structural abnormalities and aneuploidy. Hansmann (1974) gave additional data for Cytoxan treated female mice and reported 5.9 per cent of 136 metaphase II preparations were hyperploid.

Purine and pyrimidine antimetabolites

Other mutagens have also been reported to be trisomigens although, once again, the data are restricted. The pyrimidine analogue, 5-fluoro-deoxyuridine (FUdR) was tested on *Drosophila* by Traut (1978) and found to induce both chromosome loss and chromosome gain. Increases were statistically highly significant for late broods, and Traut concluded that, providing the timing of oogenesis was unaltered by the treatment, oocytes which were in interphase when treated were the most susceptible.

The folic acid antagonist amethopterin (methotrexate) interferes with nucleotide synthesis, and has been shown to produce chromosome breaks (Ryan *et al.* 1965). Röhrborn and Hansmann (1971) and Hansmann (1974) tested the effects of this antagonist on mouse oocytes and reported a very high frequency of chromosome aberrations and aneuploidy.

6-Mercaptopurine (6MCP) is a chemical mutagen which was chosen for aneuploidy testing in the mouse by Brook (1983). The choice of this particular compound was made following a report by Cacheiro and Generoso (1975) that among 615 sterile F_1 sons sired by MCP-treated males, three XYY exceptions were recovered. Brook (1983), however, was unable to find any evidence, at least for first meiotic division nondisjunction, in 6MCP treated oocytes or spermatocytes. Tests on early cleavage embryos were, however, not carried out. These could have also detected errors arising at the second meiotic division in the male, the most likely time for the YY sperm complement to have originated.

Ethylene glycol

Ostergren (1944) reported that ethylene glycol induced extreme stickiness of chromosomes in *Allium cepa*. Bhattacharya (1949) looked to see if this stickiness would lead to the induction of nondisjunction. He monitored the frequency of sex-chromosome exceptions in *Drosophila* but could detect no increases in aneuploidy. Maguire (1974) on the other hand, did report aneuploidy induction in maize. Abnormal chromosome behaviour was noted in various maize plants, and, in an intriguing scientific detective story, the origin of the abnormalities was traced to cards which were used to label the plants. Ethylene oxide had been used during the manufacture of these cards and, since ethylene oxide is known to react with water to generate ethylene glycol, this latter substance was tested on plants, with positive results. The trisomigenic activity of ethylene glycol is noteworthy because of its mode of action. Treatment with ethylene glycol induced equational division of the chromosomes at the first meiotic division (p. 36). The resulting single chromatids assorted independently at the second division, generating aneuploidy.

There are, however, not enough data available to decide whether the different responses of organisms to ethylene glycol treatment reflect a basic difference between plants and animals. More tests are required to establish this point.

Caffeine

Several workers have studied aneuploidy induction in *Drosophila* using caffeine and with varying results (Kihlman and Levan 1949; Ostertag and Haake 1966; Mittler *et al.* 1967a,b; Clark and Clark 1968; Bateman 1969; Zettle and Murnik 1973). The protocol has varied from experiment to experiment which makes comparisons difficult (Bateman 1969), but there appears to be general agreement that caffeine can significantly increase the frequency of chromosome loss. There is less agreement on whether chromosome gains occur after caffeine treatment. Mittler *et al.* (1967a,b) claimed a significant increase, Ostertag and Haake (1966) a significant decrease; for others there was no significant effect of treatment, which led Bateman (1969) to the view that no firm conclusion was possible. It would seem, however, that induction of chromosome gains has proved a much more difficult thing to demonstrate with caffeine than the induction of chromosome loss.

The observation by Kihlman and Levan (1949) that caffeine induces chromosome breaks in plants, and their observation that caffeine induces chromosome loss, but not gain, in *Drosophila*, would suggest that caffeine acts on the chromosomes. Changes to the chromosomes were reported following treatment including a peculiar form of stickiness where chromosomes were sticky only at discrete points along their length (Kihlman and Levan 1949).

However, caffeine is also known to affect other aspects of cell division, at least in plants. Mangenot and Carpentier (1944) reported that caffeine could induce binucleate cells both in *Allium* and *Triticum*, and Kihlman and Levan (1949) suggested that cell wall formation was affected to produce this condition in *Allium*. Gonzales (1967) treated *Allium cepa* and reported the formation of multipolar spindles at mitosis. Pena *et al.* (1981) treated meiotic cells of rye with 1 per cent caffeine. Treatment inhibited cytokinesis resulting in binucleate cells in which each nucleus later went through a meiotic division. Since it is known that multipolar spindles can lead to aneuploid cells which have specifically lost chromosomes (see p. 45) it might be that caffeine also induces chromosome loss through an effect on the division spindle. Nath and Rebhun (1976) suggested that the effects on cell division of caffeine and other methylxanthines were mediated by disturbing the sulphydryl-disulphide state of the spindle. The mode of action of caffeine in inducing aneuploidy may therefore be more closely related to the compounds to be described in a later section of this chapter.

Quinones

Hydroquinone and naphthoquinone are known mitotic poisons and quinones in general have been shown to fragment *Allium* chromosomes (Levan and Tjio 1948). Like caffeine, the action of quinones on chromosomes is thought to be indirect (Auerbach 1976). Parmentier and Dustin (1948, 1953) reported scattered groups of chromosomes in mouse cells following hydroquinone injection. The mitotic metaphases were unusual in that two polar groups of chromosomes were seen, together with a group of larger bodies on the metaphase plate. The latter were thought, by Levan and Tjio (1948), to be fragments but Parmentier and Dustin (1953) disagreed. They argued that they were whole chromosomes because under certain circumstances they could move to the poles. They suggested that hydroquinone was a mitotic poison affecting pre-metaphase movement.

Whatever the mechanism it is likely that hydroquinone, like caffeine, can induce chromosome loss either by acting in a radiomimetic way or by affecting chromosome movement.

COMPOUNDS ACTING ON TUBULIN AND MICROTUBULES

Colchicine and its derivatives

The best known and most extensively studied compounds acting on cell division proteins are undoubtedly colchicine and its derivatives. A detailed examination of these compounds is justified because their interaction with the spindle proteins has been extensively studied *in vitro*, and so there are interesting correlations between the action of the drugs *in vivo* and their ability to bind to tubulin—a major component of the microtubules.

It is well known that colchicine suppresses second polar body formation by its destructive effect on the spindle apparatus, thereby inducing triploids in the mouse (Edwards 1954, 1958; McGaughey and Chang 1969) and the rat (Piko and Bomsel-Helmreich 1960). In some of these studies, it was said that aneuploidy was also induced, although the chromosome techniques were not adequate to obtain accurate counts of chromosomes. Von Gelei and Csik (1939), Traut and Scheid (1974), Ramel and Magnusson (1979), and Held (1982) have, however, reported increases in sex chromosome nondisjunction in *Drosophila* following colchicine and colcemid treatment. Moreover, colcemid has been clearly shown to induce aneuploidy in mammalian somatic cells *in vitro*, owing to its deleterious effects on the mitotic spindle (Rao and Engelberg 1967; Cox and Puck 1969; Cox 1973; Kato and Yoshida 1970, 1971). An *in vivo* effect of colchicine in humans has also been claimed. Ferreira and Buoniconti (1968) studied cultures of blood lymphocytes from three gout patients under colchicine treatment and from healthy controls of the same sex and age and found, in the patients, a significant increase of cells with abnormal numbers of chromosomes, both tetraploid and quasi-diploid. The finding of an increased frequency of cells with 47 chromosomes in the blood of each of the three patients with gout suggested to the authors that their patients could have been at increased risk of producing trisomic offspring. In fact, one was reported to have come to the laboratory to submit his abnormal child to cytogenetic study. The child proved to be positive for Down syndrome, trisomy 21. However, subsequent discussion, provoked by the claims made by Ferreira and Buoniconti (1968), revealed a scepticism among other workers. Walker (1969) believed that further studies both on mitotic and meiotic tissues were required before a definitive answer could be given. Timson (1969) suggested that the apparent association between colchicine treatment for gout and having a child with Down syndrome could well be due to the relatively high probability of the two events occurring at the same time, gout and the risk of producing a trisomic child being both more marked in older individuals. Hoefnagel (1969) considered that the genetic make-up of the gouty individual, rather than the colchicine, could be the important factor predisposing to nondisjunction. The genomic mutagenicity of colchicine in mammalian meiotic cells has thus remained unconfirmed until quite recently (Sugawara and Mikamo 1980). These authors administered colchicine to female Chinese hamsters at a carefully chosen dose that did not completely arrest the formation of the first meiotic spindle in oocytes but showed a remarkable ability to induce nondisjunction. A single dose of 3µg of colchicine per gram body weight was administered at a time when spindle formation was taking place. In a total of 2124 oocytes analysed at metaphase II, following natural ovulation, the overall incidence of aneuploidy increased impressively, from 2 per cent (35/1742) in the controls to 25.9 per cent (99/382) in the experimental group. Both anaphase lagging and nondisjunction were observed. The authors believed that the abnormal conditions induced in the spindle microtubules by the *in vivo* action of colchicine was comparable to that occurring in preovulatory aging oocytes. The effect, it was suggested, was on the polymerization of tubulin, a constituent protein of the microtubules. It has been shown that meiotic nondisjunction is associated with a defect of spindle fibres (microtubules) in the first and second divisions of *Xenopus laevis* eggs aged by pre- and post-ovulatory overripeness (Mikamo 1968). Similar evidence also exists for the rat (Fugo and Butcher 1971; Mikamo and Hamaguchi 1975; Kamiguchi *et al.* 1979). The microtubule defect in preovulatory overripeness has been interpreted as inhibition of tubulin polymerization caused by the unusually prolonged retention of eggs at the germinal vesicle stage within mature follicles (Sugawara and Mikamo 1980). Polymerization of tubulin has been shown to be inhibited by colchicine in sea urchin eggs, thus blocking the formation of the spindle (Wang *et al.* 1977).

Positive results following colchicine treatment have not always been obtained, however. For example, Stavrovskaya and Kopnin (1975) reported aneuploid induction in tumour cells but there was no induction in mitotically dividing normal cells, although there was apparently no difference between the cell types in the amount of colchicine taken up or in the colchicine-binding properties of the cell components. None the less, there seems, on the whole, to be good evidence for aneuploid induction by colchicine in both mitotic and meiotic cells from higher eukaryotes.

The same can not be said for aneuploid induction in lower eukaryotes where colchicine appears to be ineffective in inducing metaphase arrest, polyploidy, or aneuploidy. This was first reported by Richards (1938) in yeast and then by Sansome and Bannon (1946) in *Penicillium* and has subsequently been confirmed by many workers. Haber *et al.* (1972) showed that colchicine was ineffective at inhibiting growth of *Saccharomyces*; colcemid was more effective, but only at very high concentrations compared to those necessary to inhibit cells of higher eukaryotes. Williams (1980) induced mitotic arrest in a proportion of dividing cells in the slime mold *Polysphondylium pallidum* using colchicine, but again only at high, millimolar, concentrations of the drug.

This dissimilarity between lower and higher eukaryotes does not reside in any difficulties of uptake by fungal cells, nor in an ability of fungi and slime molds to metabolise colchicine to a non-effective derivative. Rather it has been convincingly demonstrated that there is a basic difference in the affinity of colchicine for lower and higher eukaryote tubulin. Borisy and Taylor (1967*a*,*b*) showed that colchicine was capable of binding to the mitotic apparatus of sea urchin eggs, and there is no doubt that the major effect of colchicine is brought about by this means. Colchicine brings about the disassembly of microtubules by binding to the basic structural unit, the tubulin dimer. The complex so formed then binds to the ends of microtubules preventing the addition of further tubulin components (Margolis and Wilson 1981). Not all microtubules in a cell are equally sensitive to the effects of colchicine. The precise effect of treatment may depend on the stability of the microtubule type, so that while the microtubules of cilia and flagella are unaffected by colchicine, those in the cytoplasm may undergo depolymerization (Stebbing and Hyams 1979). The important discovery by Weisenberg (1972) that a low calcium ion concentration is a necessary condition for *in vitro* polymerization of microtubules from tubulin, has led to extensive analysis of the binding properties of several mitotic inhibitors. Bryan (1972) demonstrated three classes of binding sites in microtubules. One site is occupied by guanosine dinucleotides; a second by the *Vinca*-derived alkaloids, vinblastine and vincristine; and a third site binds colchicine, colcemid, and podophyllotoxin. This latter site is apparently different in lower eukaryote microtubules. Heath (1975) tested microtubules from the fungus *Saprolegnia ferax* and found colchicine and colcemid ineffective in disassembling microtubules.

It seems therefore that there is a basic difference between the tubulin component of cytoplasmic microtubules of fungi and slime molds on the one hand and higher eukaryotes on the other. However, not all drugs discriminate the colchicine binding site of the two types; methyl benzimidazole carbamate (MBC), for example, binds to microtubules from both sources and there is evidence that MBC and colchicine compete for the same site in higher eukaryotes (Hoebeke *et al.* 1976) (see below).

Although the major effect of colchicine is undoubtedly on the assembly and disassembly of microtubules, there are other effects of the drug, some of which may be responsible, in a minor way, for aneuploidy induction. Colchicine, for example, is known to reduce chiasma frequency and bring about increases in numbers of univalents (Levan 1939*b*; Driscoll *et al.* 1967; Dover and Riley 1973; Shepard *et al.* 1974; Pena and Puertas 1978). It has been suggested that colchicine affects a pre-meiotic control of pairing because treatments applied during or immediately after the mitosis preceding meiosis appear to be effective in reducing chromosome pairing (Driscoll *et al.* 1967; Dover and Riley 1973). The observation by Driscoll and Darvey (1970) that chiasma frequency in isochromosomes was unaffected by colchicine treatment suggested to them that the effect was on a prealignment of the chromosomes necessary for the start of the pairing process, rather than on pairing itself.

Dover and Riley (1973) also observed that colchicine treatment resulted in the formation of multipolar spindles, and these too can lead to the formation of aneuploid products.

Methyl benzimidazole carbamate and its derivatives

Although not so extensively studied as colchicine, methyl benzimidazole carbamate (MBC), which is the active breakdown product of the fungicide Benlate (Clemon and Sisler 1969), has been reported to induce mitotic arrest and aneuploidy in a variety of organisms.

Mitotic arrest in lower eukaryotes such as slime molds, has been reported by Mir and Wright (1978), Cappuccinelli *et al.* (1979), Williams (1980), and Quinlan *et al.* (1981). Benzimidazole carbamates also affect cell division in higher eukaryotes. Went (1966) reported that cell cleavage in the sand dollar, *Dendraster excentricus*, was inhibited, but there was no effect on centriole division so that cells with multipolar spindles were formed. Styles and Garner (1974) reported mitotic arrest and delay in MBC-treated rat and human cell lines. Seiler (1976) treated mouse and Chinese hamster bone marrow cells and induced lagging chromosomes and unequal chromatin masses after cell division.

Aneuploid induction by MBC has been reported in slime molds (Welker and Williams 1980) and in fungi (Hastie 1970; Kappas *et al.* 1974; Kappas 1978; Morpurgo *et al.* 1979), but the evidence for aneuploid induction in higher eukaryotes is more limited. Tates (1979) studied the effect of MBC on the frequency of aneuploid spermatids in *Microtus oeconomus* but the results were too preliminary to permit firm conclusions to be drawn. The report of unequal chromatin masses and lagging chromosomes by Seiler (1976) can be regarded as evidence of a possible induction in mammals. Seiler (1975) has evaluated the genetic risk to man of exposure to benzimidazole carbamates, including the possibility of the induction of aneuploids. The evidence that benzimidazole carbamates do induce aneuploidy, however, is so sparse that detailed consideration of these compounds would not be justified were it not for the fact that they have been shown to bind to tubulin and to interfere with the assembly of microtubules in much the same way as colchicine. Hoebeke *et al.* (1976) reported the binding of MBC derivative oncodazole, to rat brain tubulin, whilst Davidse and Flach (1977) and Quinlan *et al.* (1981) studied the binding of MBC to tubulin extracted from *Aspergillus* and *Physarium* respectively. MBC and colchicine apparently act on the same tubulin binding site, each being a competitive inhibitor of the other. They are distinguished in that the colchicine–tubulin complex dissociates very slowly, whereas the MBC derivative, parbendazole, rapidly dissociates from tubulin if the unbound drug is removed from the reaction (Quinlan *et al.* 1981). There are experiments which indicate that benzimidazole carbamates may have effects in addition to those on the spindle. Zutsch and Kaul (1975) treated the roots of bean and barley seedlings with 23 fungicides. Sixteen of these induced cytological abnormalities. Benlate induced chromatid breaks and chromosome aberrations with the highest frequency of all the tested fungicides.

Holden *et al.* (1976) reported an effect of derivatives of benzimidazole on mouse chromosomes. When injected intraperitoneally, stickiness of the chromosomes in the centromere region was induced. The results were peculiar to the mouse because no equivalent stickiness was induced in human, bat, or dog chromosomes tested *in vitro*.

Vinblastine and its derivatives

A difference at the colchicine binding site is not the only feature which distinguishes higher and lower eukaryotic tubulin. It has been shown that treatment with the *Vinca*-derived alkaloids, vinblastine and vincristine, disrupts the cell division spindle in oocytes of the annelid *Pectinaria gouldi* (Malawista *et al.* 1968), and Bryan (1972) and Owellen *et al.* (1972) have established that vinblastine binds to a specific tubulin site. The drugs are, however, ineffective against fungal microtubules (Heath 1975). The fact that vinblastine has been shown to disrupt microtubule assembly both *in vivo* (Malawista *et al.* 1968), and *in vitro* (Grisham *et al.* 1973), suggests that these drugs may, like colchicine, be effective trisomigens at concentrations which permit, albeit disturb, cell division. However, this appears to be largely untested except for a preliminary, but inconclusive, experiment by Tates (1979) using *Microtus oeconomus*. Another indication of possible trisomigenic action is the formation of tripolar spindles (Malawista *et al.* 1968).

Griseofulvin

The antifungal agent griseofulvin which is used as a treatment for fungal skin infections, has been shown by Paget and Walpole (1958) to induce mitotic arrest in rats. Malawista *et al.* (1968) demonstrated an affect on the spindle, as judged by reduced birefringence. Crackower (1972) also reported damage to the mitotic spindle in *Aspergillus nidulans*. Mitotic division is inhibited in both fungi and slime molds (Gull and Trinci 1973, 1974), although the nuclei of the plasmodial and amoebal stages of the slime molds, *Physarum polycephalim* are not equally sensitive to griseofulvin (Mir and Wright 1978). The mitotic division in the plasmodial stage is a closed one (taking place within the nuclear membrane) and has no centrioles, whilst division is open, with centrioles, in the amoebal stages.

Again, direct tests for aneuploid induction have not been extensive, but Paget and Walpole (1958) reported the formation of multipolar spindles after treatment of *Vicia faba*, the subsequent cell division resulting in the formation of irregular-sized chromatin masses which may have been aneuploid.

The evidence that griseofulvin affects microtubules is indirect, coming from observations on microtubule-dependent processes such as flagella regeneration (Margulis *et al.* 1969; Mir and Wright 1978) and the effects on the cell division spindle reported above. However, griseofulvin does not bind to tubulin dimers, nor does it prevent microtubule formation *in vitro* (Grisham *et al.* 1973; Heath 1975; Roobol *et al.* 1976). The latter workers obtained evidence that griseofulvin prevented the microtubule-associated proteins from combining with the microtubules. It has been suggested that griseofulvin disturbs a sliding function of microtubules, preventing them from moving past one another. This sliding process is a central feature of some models of chromosome movement (McIntosh *et al.* 1969).

The importance of griseofulvin in the context of drugs which interfere with chromosome movement is that it demonstrates by its action that some compounds may interfere with microtubular function in ways other than by binding to tubulin and thus preventing microtubule assembly. More compounds which might fall into this category are discussed in a later section of this chapter.

Many other compounds which bind to tubulin or to microtubules have been investigated. Some of these are interesting because they share structural features which are thought to be important in their interaction with the binding site. Harrison *et al.* (1976) and Wang *et al.* (1977) examined the effects of mescalin and stegnacin, respectively, on tubulin polymerization. They concluded that compounds with a trimethoxyben-zene ring (which both the above compounds and colchicine possess) binds to the colchicine binding site of tubulin.

The herbicide trifluralin has been shown to induce aneuploidy in *Neurospora* (Griffiths 1979): it is also a tubulin binder. Hess and Bayer (1977) demonstrated that it bound to tubulin from *Chlamydomonas*, and suggested that there was a difference between plant and animal tubulin, since trifluralin has not been shown to bind to tubulin from any animal source.

The antitumour drug, taxol, has been the subject of recent investigation and has turned out to have an interesting mode of action. Its antimitotic action results from binding to tubulin microtubules, but it has quite the opposite effect compared to microtubule inhibitors such as colchicine. Taxol promotes the formation of microtubules so that the cytoplasm of a treated cell contains microtubules whose formation is the result of spontaneous nucleation. It has been shown that taxol reduces tubulin dimer concentrations in cytoplasm (Schiff *et al.* 1979) and it has been suggested (de Brabander *et al.* 1981) that taxol displaces the equilibrium between tubulin dimers and intact microtubules. Microtubule formation is favoured by reductions in the critical concentration of tubulin.

Although some of the compounds mentioned in the above section have not been tested for their trisomigenic action, they are prime candidates to have such an effect because they are effective inhibitors of cell division. Investigation into the mode of action of tubulin binders shows that they can disrupt microtubule assembly in more than one way. They reveal differences between the structure of microtubules which raise serious doubts on the efficacy of testing methods which use organisms remote from man (see Chapter 8).

COMPOUNDS ACTING ON THE SPINDLE IN OTHER WAYS

In a selective compilation of chemical mutagens, Barthelmess (1970) has listed over 120 chemicals which have been reported to interfere with the spindle in one or more organisms. The trisomigenic effect of many of these compounds has not been determined. The compounds included in this section are therefore only a selected sample of spindle inhibitors. They have been chosen either because they have been worked with fairly extensively and may have an interesting mode of action, or because they could be an important contaminant of the environment to which at least some human individuals might be exposed.

p-Fluorophenylalanine

Whilst microtubules respond differently to various drugs this may not be the only cell division component to be susceptible to chemical treatment. Various amino-acid analogues have been shown to increase the duration of mitosis, and amongst these, the analogue *p*-fluorophenylalanine (*p*FPA) has been studied the most. *p*-FPA has been shown to be trisomigenic at both mitosis and meiosis in a wide range of fungal species (Lhoas 1961; Stromnaes 1968; Da Cunha 1970; Day and Jones 1971; Griffiths and Delange 1977; Parry 1977). In *Sordaria brevicollis* it has been shown to increase meiotic errors at both the first and second meiotic divisions (Bond and McMillan 1979). Fungi seem to be particularly susceptible to treatment with this analogue, but attempts to assay its effects in mammals have been less successful, yielding inconclusive or negative results (Tates 1979; Brook 1983). However, effects on mammalian cell division have been recorded. Biesele and Jacquez (1954) treated human cell lines with *p*FPA and induced metaphase arrest and tripolar mitoses. Sisken and Wilkes (1967) and Sisken and Iwasaki (1969) studied the effects of several analogues all of which affected the duration of mitosis. Only *p*FPA increased the duration of metaphase leaving other parts of the mitotic cycle relatively unaffected. They suggested that *p*FPA treatment resulted in the analogue being incorporated into a cell-division-specific protein. Vaughan and Steinberg (1960) demonstrated the incorporation of *p*FPA into proteins in an *in vitro* system whilst Westhead and Boyer (1961) were able to show that the analogue was incorporated in the place of phenylalanine.

The mode of action of pFPA is not firmly established and it is possible that *p*FPA interacts with tubulin in some way. It is known that α-tubulin is tyrosylated by a specific tubulin–tyrosine ligase (Raybin and Flavin 1975; 1977) and that the post-translational modification of α-tubulin by this enzyme can be modified so that some tubulin molecules have phenylalanine instead of tyrosine added to the COOH terminus (Rodriguez and Borisy 1978, 1979). It is possible, although certainly not proven, that *p*FPA could affect tubulin by affecting post-translational modification and that this in turn could affect microtubule function leading to aneuploidy.

Regardless of the mechanism of action, the important point to be emphasized is that *p*FPA is an example of a compound which is trisomigenic in lower eukaryotes but has given negative results when tested in higher eukaryotes. This again may reflect a variation in susceptibility of the spindles of different species to chemical treatment.

Heavy metals

In an extensive series of experiments in the 1940s, Levan and Ostergren found that there was, for a large number of chemically diverse substances, a correlation between the water solubility of a compound and the minimum concentration necessary for it to induce mitotic arrest (Ostergren and Levan 1943; Levan and Ostergren 1943). They suggested that for these substances, the c-mitotic activity was due to some physical property of the molecules. This correlation is represented graphically in Fig. 7.1.

Some substances were notable in that they induced arrest at much lower concentrations than would have been predicted from their water solubility. Colchicine was one such compound (Ostergren 1951), and several organic compounds containing heavy metals have also been shown to have similar effects (Onfelt and Ramel 1979). Ostergren suggested that this high sensitivity could be explained by a specific chemical interaction between such compounds and the spindle.

Mercury

Ramel (1969) investigated the effects of treating *Allium* roots with organic mercury compounds and showed that c-mitotic activity was exhibited at very low concentrations. This interaction of mercury with the spindles can lead to aneuploid formation. Sass (1937) showed that corn seedlings treated with ethylmercury phosphate contained nuclei of various sizes ranging from micronuclei to enlarged polyploid nuclei. More direct tests for aneuploidy induction have been carried out on *Drosophila* by monitoring for sex chromosome aneuploidy. After treating adults or larvae with methyl mercury hydroxide or phenylmercury acetate, Ramel and Magnusson (1969) reported small, but statistically significant, increases in the frequency of matroclinous exceptions. The interesting feature of these

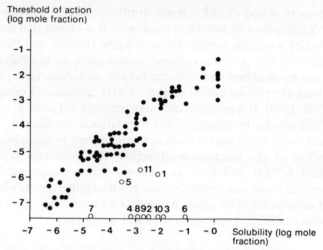

Fig. 7.1. Illustration of the correlation between the water solubility of a number of different chemical compounds and the minimum concentration necessary for the induction of mitotic arrest (closed circles). Some substances induce arrest at much lower concentrations than predicted from their water solubility (open circles): (1) colchicine; (2) methyl mercury dicyandiamide; (3) methyl mercury hydroxide; (4) phenyl mercury hydroxide; (5) mercury chloride; (6) trimethyl tin chloride; (7) tributyl tin chloride; (8) diethyl lead chloride; (9) triethyl lead chloride; (10) trimethyl lead chloride; (11) lead nitrate. (From Ostergren (1951) and Ramel and Magnusson (1979).)

results was that there were no increases in the frequency of patroclinous exceptions which would result from chromosome loss. It seems that treatment induced meiotic errors resulting in chromosome gain but, for some reason, there was no equivalent induction of chromosome loss. This specificity of the treatment disappeared if the X chromosomes were heterozygous for a chromosome inversion, when increases in both matroclinous and patroclinous exceptions occurred. Ramel and Magnusson considered several explanations for the unexpected finding that mercury induced chromosome gain only, such as the selective survival of the gametes or extra replication of the X chromosome, but were unable to draw any firm conclusion on the nature of the unusual specificity, largely because the phenomenon disappeared in the presence of inversions.

Another feature of chemically induced aneuploidy is illustrated in these experiments: increases are seldom very large. For some unknown reason a threefold increase in aneuploidy appears to be the norm. Examples of the increases which Ramel and Magnusson obtained in their experiments can be found in Table 7.1, and these appear to be characterstic of chemically induced aneuploidy.

TABLE 7.1.   *Effect of larval treatment with methylmercury hydroxide (0.25 mg Hg/l substrate) on nondisjunction in females (From Ramel and Magnusson 1979)*

Exp. no.	Parental females	Treated			Control		
		% Exceptions		Total number	% Exceptions		Total number
		XXY	XO		XXY	XO	
1	y w sn/y w sn	0.13***	0.18	100 734	0.07	0.20	113 672
2	y w sn/y w sn	0.07*	0.16	76 906	0.04	0.16	66 863
3	y w sn/y w sn	0.10*	0.16	24 061	0.05	0.14	35 320
4	y w sn/y w sn	0.08**	0.15	15 446	0.02	0.16	23 110
5	y w sn/y w sn	0.15**	0.19	34 150	0.08	0.17	36 550
6	y wᵃ f/y wᵃ f	0.06	0.02	23 250	0.04	0.04	30 950
7	y wᵃ f/y wᵃ f	0.06**	0.02	44 250	0.02	0.02	75 550
8	Berlin	0.05	0.03	44 200	0.03	0.04	55 800
9	Berlin	0.03	0.02	40 100	0.02	0.03	64 200
10	Karsnäs 60	0.04	0.02	41 100	0.04	0.02	61 600
11	Karsnäs 60	0.05	0.04	56 150	0.04	0.04	71 150

Female crossed to y w sn/sc^8 Y in experiments 1–5; crossed to y wᵃ f/y$^+$ Y Bˢ in experiments 6–11.
*p = < 0.05 > 0.01; **p = < 0.01 > 0.001; ***p = < 0.001.

With regard to the mechanism of action of mercury in inducing aneuploidy in *Drosophila*, Ramel (1969) has suggested that the c-mitotic activity could be explained by the high affinity of mercury for sulphydryl groups of the spindle. The importance of sulphydryl groups and disulphide bridges in spindle function was suggested by Mazia (1958), and subsequent work has shown that the disrupting activity of mercury on cell division in tissue culture is affected by the sulphur content of the medium. Ramel (1969) found no effect of mercury on lung tissue cultures and suggested this was because of the high sulphur content of the culture media. Fiskesjo (1970) showed that mercury would induce c-mitosis in human leukocytes provided that the concentration of cysteine and cystine in the medium was controlled. Support for the idea that heavy metals affect the balance between SH groups and S—S bonds in the spindle is provided by experiments in which the effects of sulphydryl compounds on mitosis have been studied. Ramel (1969) showed that 2, 3-dimercaptopropanol was an antagonist of the c-mitotic effect of mercury.

It is likely, therefore, that the trisomigenic effect of mercury is mediated through disruption of spindle function, but the effects may not be confined entirely to the spindle. Klasterska and Ramel (1978) showed that the chromosomes of the grasshopper, *Stethophyma grossum*, were affected by treatment with methylmercury hydroxide. Sub-chromatid aberrations, some breaks and stickiness were all induced and it has been suggested (Onfelt and Ramel 1979), that treatment affected chromosome folding. There is good evidence for the involvement of sulphydryl groups in the process of chromosome condensation (Ord and Stocken 1966).

In the mouse, when mercury was tested on dictyate oocytes (Jagiello and Lin 1973), no effects on chromosome segregation were recorded. A number of epidemiological studies in man have, however, indicated possible dangers from mercury exposure although the data are conflicting. Statistically significant increases in aneuploidy (but not aberrations), were recorded in the blood lymphocytes of exposed mercury workers by Verschaeve *et al.* (1976, 1978), but in a later study, carried out by the same authors (Verschaeve *et al.* 1979), no effects could be detected. It was speculated that measures taken to achieve protection had reduced the effects of mercury to the exposed workers.

Lead

Lead compounds have also been tested and shown to induce aneuploidy. Ahlberg *et al.* (1972) tested several organo-lead compounds and showed that they had severe effects on the spindle of *Allium cepa*. In preliminary experiments a threefold increase in chromosome loss was recorded in tests on *Drosophila* with no significant increase in gains. Later experiments, however (Ramel and Magnusson 1979), gave similar results to the organo-mercury compounds with increases only in matroclinous exceptions.

Cadmium

Cadmium chloride has been reported to induce nondisjunction in primary oocytes of the mouse (Watanabe *et al.* 1977; Watanabe and Endo 1982), and golden hamster (Watanabe *et al.* 1979). Doses of the heavy metal ranging from 1 mg/kg to 4 mg/kg were applied at the preovulatory stage of meiosis (diakinesis/metaphase I) in the golden hamster. At the highest dose, four hyperploid counts were recorded in 95 metaphase II complements analysed, giving (by doubling) an aneuploidy level of 8.4 per cent. The increase was significant ($P<0.05$).

In initial experiments in the mouse, doses of 3 mg/kg and 6 mg/kg were injected into females, also at the preovulatory stage of diakinesis/ metaphase I. In the highest treatment group, three hyperploid counts were recorded in 164 metaphase II oocytes examined. The increase in aneuploidy over control levels was, however, not significant. Subsequently, Watanabe and Endo (1982) analysed the chromosomes of blastocysts from similarly treated mice to examine the effect of cadmium chloride through the oogenesis stage to the preimplantation stage. Monosomic, trisomic, and triploid blastocysts were all observed to increase in the treated groups, with the overall incidence of anomalies being about three times higher than that found in oocytes examined at metaphase II. An increase in triploid complements was the most prominent effect of treatment, but at the dose of 3 mg/kg, three trisomics out of 63 blastocysts analysed were reported. The mechanism of action of cadmium in inducing segregational errors was not, however, investigated.

Zinc

There is no evidence for aneuploid induction by zinc, although Herich (1969) claimed that zinc was a pre-prophase poison. This conclusion was reached because zinc had no effect on the spindles of *Vicia faba* but did affect the mitotic index, perhaps by reducing the number of cells entering prophase.

Anaesthetics

Nitrous oxide has been shown to produce aneuploidy. In plants, Ferguson *et al.* (1950) reported the c-mitotic activity of various gases including nitrous oxide, and Ostergren (1954, 1957) demonstrated the production of aneuploid progeny after treatment of *Phalaris* and *Crepis*. Other inhalation anaesthetics have also been shown to produce segregation errors of the chromosomes. Grant *et al.* (1977) reported increased aneuploidy after the roots of *Vicia faba* had been exposed to halothane or methoxyflurane bubbled through water in which the roots were placed. One per cent of cells in the exposed plants were aneuploid.

Aneuploid induction occurs in animals too. Kusyk and Hsu (1976) induced errors using three halogenated anaesthetics, halothane, enflurane, and methoxyflurane both *in vitro* (on mammalian cells) and *in vivo* (on birds). The induced errors included unequal division, chromosome lagging, stickiness, and cells with multiple nuclei. There was no evidence for chromosome breakage. Also, anaesthetics such as ether, chloroform, nitrous oxide, etc. have been found to cause mitotic anomalies in cultured HeLa cells, the microtubules being disarranged and the cells arrested at metaphase (Brinkley and Rao 1973).

Aneuploidy induction at meiosis has also been observed. Pathak and Hsu (1977) monitored the effects of halothane injected into the testes of Chinese hamsters. They reported an increase in aneuploids when second meiotic metaphases were scored but gave no data. Kaufmann (1977), in the mouse, showed that anaesthesia administered shortly before ovulation could interfere with meiotic chromosome segregation, leading to aneuploid eggs and embryonic mortality. Clements and Todd (1981) recorded increases in aneuploidy in *Drosophila* using the isochromosome method of Clark and Sobels (1973). Adult flies were fed solutions of halothane in sucrose which resulted in small increases in aneuploid progeny. Allison *et al.* (1970) have given an indication of how these inhalation anaesthetics might bring about their effect on chromosome segregation. They studied the effects of anaesthetics on the microtubules of *Actinosphaerium nucleofilum*, using measurement of axopod collapse to indicate the extent of microtubules by analogy with the effects of colchicine. When Brinkley and Rao (1973) studied HeLa cells exposed to nitrous oxide, the cells showed typical colchicine-like metaphases but, on examination, both interpolar and kinetochore microtubules were present, so the gas did not seem to exert its effect through complete disruption of the microtubules. There was some effect on the microtubules, however, because they did show atypical alignment. Prolonged arrest gave tripolar cells.

In these experiments and those of Ferguson *et al.*, and Ostergren, the material was exposed to the gas under high pressure, but Brinkley and Rao (1973) showed that neither pressure itself nor anoxia were responsible for the induction because nitrogen at a similar pressure had no effect.

It was suggested, by these latter authors, that most, if not all, anaesthetics share some cellular organelles as their action targets: the spindle and possibly also the centrioles. Induced meiotic errors in man could contribute to the higher levels of fetal mortality and/or congenitally abnormal progeny reported among individuals or their spouses exposed in operating theatres to the effects of anaesthetic gases (Cohen *et al.* 1971; Knill-Jones 1975; Corbett *et al.* 1974; Pharoah *et al.* 1977).

Hormones

The current interest in hormones, or 'hormonal imbalance', as a possible explanation of the maternal-age effect in man, makes the studies which have been carried out on aneuploid induction by hormones very topical.

It is known that before women, and most female rodents, cease cycling, they undergo a period of irregular cyclicity during which an increased incidence of prolonged cycles occurs (Sherman *et al.* 1976; Nelson *et al.* 1981). The possible hormonal basis for such extended cycles in aging C57Bl female mice has been shown, by Nelson *et al.* (1981), to be a striking impairment in the preovulatory rise in plasma oestradiol which normally is required to trigger the LH surge at ovulation. In 10–12.5-month-old C57Bl/6J mice, the preovulatory increase of plasma oestradiol was found to be delayed by about one day, a period corresponding to the average net increase in cycle length at this age. The results are consonant with the finding that plasma oestradiol is also markedly reduced in premenopausal women and associated with irregular cyclicity (Sherman *et al.* 1976).

It is known that abortions and fetal anomalies occur more frequently in pregnancies of older women (Hendricks 1955), and it has been shown in the rat (Fugo and Butcher 1971) that it is the prolonged cycles of older females that are important in producing such anomalies. Fugo and Butcher (1971) have suggested that the prolonged cycles of older women may produce overripe ova which, when fertilized, are less capable of under-going normal embryologic development. In an earlier section, it was suggested that such overripeness of ova was associated with spindle degeneration and this in turn led to nondisjunction. An alternative hypothesis, in which it is postualated that the delay in meiosis caused by overripeness brings about an asynchrony between the timing of chiasma terminalization and chromosomal alignment on the spindle, has been put forward by Crowley *et al.* (1979).

If the interval from resumption of meiosis to ovulation lengthens in older women, bivalents, which become attached to spindle fibres roughly 18 hours after resumption of meiosis in 20-year-old women, will, on the model, predictably be delayed in attachment as the oocyte ages. There will thus be an increasing chance of losing chiasmata before proper alignment on the spindle has been established. Since meiotic bivalents are held together by their chiasmata against strong mutual repulsion during late diplotene and diakinesis, smaller bivalents with fewer chiasmata will be the most vulnerable to premature separation during terminalization. This is a view consistent with the observation in human spontaneous abortions that the smaller chromosome trisomics (with the exception of trisomy 16) are more strongly age-related than the large chromosome trisomics (Hassold *et al.* 1980*a*).

The model is invoked by Crowley *et al.* (1979) to explain the higher incidence of Down syndrome in both aged and very young women (Erickson 1978). They suspect that this may result from a higher variability of hormone levels among cycles as the complex neuroendocrine feedback controls are first being established in the post-menarche period and are then becoming disturbed again at the menopause (Treloar *et al.* 1967).

As Crowley *et al.* (1979) point out, the hypothesis yields some testable predictions one of which is that the incidence of trisomy will respond to experimental manipulation of intrafollicular hormone levels. Indeed, a number of studies in which the influence of hormones on meiotic maturation in general, and on the production of chromosomal anomalies in particular, have been tested, are to be found in the literature. Moor (1978) for example, has shown that steroids are essential for *in vitro* ovine oocyte maturation compatible with subsequent normal embryonic development to full term. Inadequate steroid support during oocyte maturation (oocytes cultured within their follicles), results in anomalies in fertilization, delayed cleavage and an almost total failure to undergo differentiation and blastocyst formation. The experiments of McGaughey (1977) also suggest that both progesterone and oestradiol are essential for normal oocyte development in the pig. Chromosomal examination of air-dried oocytes indicated that intact and denuded oocytes cultured in either progesterone or oestradiol-17β exhibited an increased incidence of diploidy and hyperploidy at metaphase II. Those cultured in media containing both progesterone *and* oestradiol showed a higher incidence of normal haploidy than did control oocytes. Dose-related changes in chromosomes of the dog testes treated *in vivo* with progesterone have been reported by Williams *et al.* (1971) and in a subsequent series of experiments (Williams *et al.* 1972), the same authors showed that stickiness, clumping, and condensation of oocyte diplotene figures occurred in progesterone-injected Chinese hamster females.

A meiotic study of the *in vitro* and *in vivo* effects of components of currently used oral contraceptives on the oocytes of six female mammalian species was made by Jagiello and Lin (1972). *In vitro* inhibition of division and atresia was found in several species with both the oestrogens and progestogens, as well as a low fequency of chromosome breaks and rearrangements, anaphase lags, spindle abnormalities and polar body suppression. A clear extrapolation for a direct action of oral contraceptives on human oocytes was not, however, claimed by the authors. An increase in aneuploidy in mouse oocytes analysed at metaphase II following administration of a gestagen used as oral contraceptive has, however, been claimed by Röhrborn and Hansmann (1974), but the effect was only observed at the highest dose used.

The possible influence of endogenous hormones on nondisjunction is also suggested by studies on somatic cells cultured *in vitro*. Mauer (1963) reported a reduced mitotic index, that is, an inhibition of cell division, of human leukocytes treated with testosterone, progesterone, or chorionic gonadotrophin, but at hormone levels which were ten times the normal circulating levels. Cultures treated with oestradiol and hydrocortisone had a higher frequency of cells with a non-modal chromosome number. Rao and Engelberg (1967) observed that aneuploidy developed in HeLa cells cultured in the presence of certain steroid hormones, but only the oestrogens with OH groups at the 3rd and 17th positions of the steroid skeleton appeared to be effective in inducing the anomaly. Diethylstilbestrol (DES), a synthetic oestrogen, was also effective, a claim which has also been made by Chrisman (1974) for DES-treated embryos of the mouse and by Chrisman and Hinkle (1974) for mouse bone marrow cultures. Testosterone and hydrocortisone acetate were, however, ineffective. Agrell (1954) has also observed that oestradiol, at concentrations of $1-4 \times 10^{-5}$M, induces in the cells of sea urchin embryos chromosome bridges, detachment of chromosomes from the spindle and apolar spindles. He noted no real c-mitotic effects, a factor indicative of the essentially different nature of the effects of oestradiol and colchicine.

Oestradiol in high concentrations has also been shown to produce aneuploidy in cultured human synovial cells of some individuals but not others (Lycette *et al.* 1970). These authors suggested that the potential of hormones as nondisjunctive agents should be further investigated. There still remains, however, a dearth of information on the role of hormones and hormonal imbalance on aneuploidy induction and on the possible mode of action both in relation to natural aging in women and in relation to the use of oral contraceptives in younger women. Harlap *et al.* (1980) have studied the incidence of spontaneous abortion in women who stopped using oral contraceptives in the few months prior to conception, as well as those who became pregnant while actually on the pill. Conceptions occurring immediately after stopping the pill were followed by a small but non-significant increase in spontaneous abortions. After oral contraceptive failure, there was also a small but non-significant increase in first (but not second) trimester losses. Whether it was known if any of the losses were due to aneuploidy among the fetuses was not, however, stated.

Acenaphthene

Acenaphthene is an insecticide and a member of the naphthalene series used by Levan and Ostergren (1943) when they established the relationship between the solubility of a compound in water and its c-mitotic activity. It was earlier reported by Kostoff (1938) to increase aneuploidy at both mitosis and meiosis in the seedlings of various grasses and the floral buds of *Nicotiana*. The induction was brought about through the absence of chromosome movement which resulted in the formation of nuclei of irregular size. Pathak and Hsu (1977) reported induction when acenaphthene was injected into the testes of Chinese hamsters.

Apart from the reports of its c-mitotic activity little is known about the mechanism of aneuploid induction by acenaphthene. It is not a compound to which the spindle is exceptionally sensitive, and it may be, as suggested by Levan and Ostergren, that aneuploidy induction by acenaphthene simply reflects a general physical property of substances to interfere with the spindle when it is exposed to concentrations above a threshold value.

Hexachlorocyclohexan

This compound is also an insecticide, and one which possibly illustrates another mode of action for chemicals interfering with the spindle. Nybom and Knutson (1947) tested the c-mitotic activity of four isomers of hexachlorocyclohexan (HCCH). γ-HCCH induced full c-mitosis, whereas the other isomers induced partial c-mitosis or had no c-mitotic activity at all. The γ-HCCH isomer does induce aneuploidy in plants (Kostoff 1948; Giménez-Martín and López-Sáez 1974). The latter workers reported the frequent formation of tripolar and multipolar cells which generated aneuploid daughter cells. It might be therefore that this compound induces aneuploidy by a means other than the induction of partial c-mitosis.

CONCLUSIONS AND FUTURE WORK

Data on chemically induced aneuploidy are unsatisfactory in several respects:

1. Much of the evidence for aneuploid induction is indirect, coming from observations on cell division where treatment has led to chromosome lagging or to unequal size of nuclei.

2. Induction of aneuploidy is often recorded only from studies on mitosis. Tests on meiosis in higher organisms are more time-consuming and expensive, yet compounds specifically affecting meiosis (for example by interfering with chromosome pairing) may exist and these would go undetected if reliance was placed solely on mitotic systems.

3. When a compound has been tested in more than one organism, it has sometimes given positive results in one case and negative in another.

4. There is even heterogeneity of response when the same chemical is tested in the same organism but by different workers, e.g. caffeine in *Drosophila*. Griffiths (1979) has discussed the problem of heterogeneity of response to chemical induction in a *Neurospora* test system.

5. Chemical induction is seldom very great. With the exception of colchicine in higher organisms and *p*FPA in fungi, there are no examples of chemicals which have given consistent and large increases in aneuploidy over a range of testing methods.

Perhaps the most worrying aspect of all, however, is the strong indication that the various testing methods do not all reveal sensitivity to the same chemicals. Lower eukaryotes, for example, are insensitive to colchicine; higher eukaryotes to induction by *p*FPA. This is unfortunate because the tests using microbial eukaryotes are cheap and convenient compared with those using mammals. It would seem, however, that the results of tests with microbial systems cannot be extrapolated, with any degree of confidence, to higher organisms. On the other hand, the routine use of mammals for meiotic aneuploidy testing probably cannot be considered a practical proposition for screening the many chemicals to which man is exposed. Considering the range of available test procedures, there does not appear to be one which can be adopted with complete confidence, and which will detect all compounds affecting chromosome segregation.

8. Conclusions and future perspectives

The aim of this final chapter will be to focus the reader's attention on the outstanding problems which, in our opinion, remain to be solved in aneuploidy research, highlighting topics in need of further consideration and attempting to pinpoint areas requiring more extensive investigation.

In the course of writing the book, we, as authors, have been impressed by the large body of experimental aneuploidy data which has already been amassed in various organisms. At the same time, however, we have been surprised by just how little is really understood about the mechanisms involved in aneuploidy production. Investigations in organisms like *Drosophila* and fungi, which permit a fairly complete analysis, justify the opinion that there is more than one cellular route to aneuploidy. The responsible inducing agent, whether it be ionizing radiation, a chemical pollutant, temperature variation, or a spindle poison appears mainly to determine the initial type of lesion. Breakage phenomena characterize much (perhaps all) radiation-induced aneuploidy and may also be a feature of the aneuploidy induced by chemical agents of the type which mimic the effects of radiation. On the other hand, direct effects on the spindle microtubules or the kinetochores seem to characterize other forms of aneuploidy, for example those attributable to heat or cold shock. Temperature variation may thus mimic the effects of known spindle poisons and inhibitors like colchicine, bringing about direct effects on chromosome segregation. In addition, chromosome pairing may be affected, resulting in aneuploidy as a secondary consequence should the chromosomes fail to segregate normally.

It is important to know the relative contributions made by one particular form of error or another, and even more importantly, to ascertain to what extent extrapolations can be made from organism to organism. Although meiosis appears to be, in evolutionary terms, a conserved biological process (Hotta *et al.* 1977), the immediate environment in which the meiotic division takes place varies enormously from one species to another. It is therefore possible that the relative frequency of each specific defect will also differ widely. We have seen in the foregoing chapters how differences exist in the structure of microtubules between lower and higher eukaryotes, and this raises serious doubts about the efficacy or relevance of testing methods aimed at assessing potential human aneuploidy induction, using organisms, like fungi, which are remote from man. The very fact that mammals are able to detoxify in the liver or exclude from the testis (blood-testis barrier) many noxious compounds which a fungus or fruitfly could not, raises another question of suitability of some of the screening systems employing lower eukaryotic organisms. A particular compound which might be cited in this respect is the amino-acid analogue *p*-fluorophenylalanine (*p*FPA) which, in fungi, gives a strongly positive response in terms of aneuploidy induction but which, in the mouse, has been found to give negative results. Were all aneuploidy testing to be carried out in a higher eukaryote like the mouse, and on germ cells, no doubt a closer degree of reliance could be placed on the outcome of the tests and a more relevant extrapolation be made to the situation in man. Screening systems using mammalian germ cells are, however, tedious and time-consuming and cannot possibly be applied in all cases where potential danger from an environmental trisomigen might exist. Only selected compounds therefore could probably be tested. Moreover, the mouse appears, in general, to show a much lower propensity for spontaneous (and perhaps induced) nondisjunction and a less marked maternal-age effect than man. Selection of an appropriate strain (such as CBA) is therefore important, in order to minimize these differences.

Much remains to be discovered about why different species show such marked variations in spontaneous aneuploidy levels and why, within any one species, differences exist in the degree to which individual pairs of chromosomes undergo nondisjunction. An explanation, for example, is still awaited for the exceptionally high level of trisomy 16 among human spontaneous abortions. Basic differences which may influence aneuploidy induction from one species to another, or influence the levels of nondisjunction between individual chromosome pairs might include the order and arrangement of chromosomes in the genome (Bennett 1982), their size and nature of attachment to the spindle (Ford and Lester 1982), or their relationships with nucleoli (Mirre *et al.* 1980).

Further investigation into the chromosomal complements of human spermatozoa along the lines already so successfully begun by Rudak *et al.* (1978) and Martin *et al.* (1982) would provide us with insights into the aneuploidy level of the male gametes. This should give us a clearer view of the relative paternal contribution made to the apparently high level of aneuploidy at conception in man. Moreover, knowledge would be gained of the relative rates with which individual chromosome pairs in the male germ line undergo nondisjunction. The technique, if applied to males in exposed populations, could also be used to estimate or monitor increases in aneuploidy brought about by environmental or occupational mutagenic compounds. Such information, unfortunately, is much more difficult to obtain in the human female on account of the enormous difficulties encountered in obtaining ovulated ova for chromosomal analysis. One possible way to determine aneuploidy levels in very early embryonic human material would be to repeat the type of study made by Hertig *et al.* (1965) and to analyse chromosomally the early products of conception in women who have had their uteri and Fallopian tubes removed.

Another area requiring intensive study is that concerned with the role of the maternal environment in aneuploidy induction in general, and in the age-related increases in particular. If prior events in the fetal ovary are not important, then how can conditions in the adult female influence events at the first meiotic division in the oocyte and in particular, in the aged female? Some suggestions have already been put forward such as incomplete spindle formation (Sugawara and Mikamo 1983), poorer physiological conditions in the follicle of older females (de Boer and van der Hoeven 1980), hormonal imbalance (Lyon and Hawker 1973), and so on. So far, however, experimental studies into these possibilities have only just been touched upon. Maternal aging remains the overwhelming aetiological factor in human aneuploidy, yet how little we understand about the mechanisms involved. In our opinion, the maternal-age effect on aneuploidy in man remains one of the major unsolved problems of human cytogenetics. Were it to be found that some factor in the maternal environment were the responsible underlying cause of age-related aneuploidy, then prophylactic measures could perhaps be taken to avoid aneuploid conceptions.

How concerned should we also be with the monitoring of human populations to prevent aneuploid births which are not related to maternal aging? It has been seen how agents such as ionizing radiation can induce aneuploidy in a variety of species, and even the low doses to which human beings are generally only exposed, may be harmful. The awareness of the dangers of ionizing radiation for many forms of damage has already led to a heightened concern for the protection of the public and individual alike. Increasingly now, concern is also being expressed at the potential dangers of many chemical compounds to which workers and populations are exposed. Much remains to be done on the part of governments and employers in providing protection against these, and the need for vigilance is great: mutagen screening will probably continue for many years to come using a range of tests and end-points. Primarily because of the cost and tedium of the work involved, screening for numerical anomalies has been badly neglected in environmental mutagenesis in the past. Some consolation, however, can be gained from the experimental studies carried out to date among mammalian species: these indicate that chemical induction of aneuploidy is seldom great. Perhaps at least in the male, the new technique of hamster egg/human sperm fusion may offer a means of testing for environmental hazards among large exposed populations of workers.

By the increased use of antenatal screening, a greater proportion of human aneuploid embryos could be aborted in the future, and the social and financial burden on society alleviated to some extent. To screen all pregnant women could see the total elimination of the trisomy 21 condition and the prevention of many other chromosomally abnormal births. Screening of women over 35 years could lead to the prevention of about one-third of cases. However, when financial resources and skilled manpower are scarce, all-embracing comprehensive screening programmes may not be feasible. Society must then decide how to use available resources to best advantage and choose where to draw the line between women to be screened and those to be excluded from the programme. Tremendous progress has been made in recent years in techniques of screening for congenital malformations and the interest with which this facility has been received by clinicians has grown steadily. Much interest has also been aroused in the cost-effectiveness of the screening programmes and, in the hope of maximizing the benefits, consideration has been given to the prevention of Down syndrome and other aneuploid births (Stein *et al.* 1973; Hagard and Carter 1976; Mikkelsen *et al.* 1978). For Scotland alone, there are likely to be around 2000 affected trisomy 21 individuals, making Down syndrome the predominant single cause of severe subnormality (Brotherston 1978).

The economic benefit of preventing the birth of a Down syndrome individual is the cost to the community of his or her care. In the case of an abortion followed by a successful normal pregnancy (replacement), this is the difference between the cost of caring for a handicapped person and that of caring for an average person. When there is no further pregnancy (no replacement), the cost is the total costs of caring for a handicapped person (Hagard and Carter 1976). In estimating these costs, an assessment of the use of resources by a nominal cohort of 100 people liveborn with Down syndrome was made by Hagard and Carter (1976). In both the replacement and non-replacement situations, the economic benefit for women over 40 years exceeded the costs. Corresponding calculations for women aged 35 years and over showed costs roughly to equal benefits; for women under 35 years, the economic benefits were less than costs. Similar assessments to these have been made in Denmark by Mikkelsen *et al.* (1978) and in British Columbia (Canada) by Sadovnick and Baird (1981). Prenatal diagnosis in both countries was cost beneficial for the maternal-age group over 35 years and gave a considerable economic benefit for the age group over 40. In New York City too, the cost of screening mothers over 30 years has been found to be less than that of caring for cases of Down syndrome among them (Stein *et al.* 1973).

The cut-off point in age of mothers being referred for antenatal screening might in future be lowered as resources permit, and if the mean age of mothers of Down syndrome babies is coming down, as current data would indicate, serious consideration must obviously be given to this possibility. The costs of non-intervention are certainly considerable both in economic terms and in human misery. Mikkelsen *et al.* (1978) have pointed out that it would be of immense value to identify families with a genetic predisposition to aneuploidy as well as women at risk because of their maternal age. They believe that intensive research should be encouraged in this direction. The suggestion of Hecht (1982) that sex chromosome mosaicism in the blood might be an indicator of an increased risk of meiotic nondisjunction is one which perhaps should be considered further.

The problem of Down syndrome and society's response to it cannot, however, rest solely on consideration of costs and benefits. Intangible aspects such as anxiety and anguish on the part of the families must also be considered. Induced abortion, used as a means of controlling such births, may not be considered acceptable by some sections of the population or within certain ethnic groups. For these women in particular, but also for women and society in general, the possibility of a prophylactic measure to prevent aneuploidy from arising in the first place, would surely be considered infinitely more desirable than termination of a pregnancy.

As Hagard and Carter (1976) have stated in the concluding paragraph of their paper, 'If Down's syndrome is socially unacceptable, provision of a programme to reduce its birth prevalence by scarcely a third would be an inadequate response. Conversely, failure to implement a programme for all maternal age groups would imply that there were other, perhaps more appropriate, responses to the problem of Down's syndrome. Since this would call into question any programme directed at identification and termination of affected pregnancies, it would be logical to resolve this dilemma before any programme was started.'

References

Agrell, I. (1954). Oestradiol and testosterone propionate as mitotic inhibitors during embryogenesis. *Nature, Lond.* **173**, 172.

Ahlberg, J., Ramel, C., and Wachtmeister, C.A. (1972). Organo lead compounds shown to be genetically active. *Ambio* **1**, 29–31.

Alberman, E.D. (1978). Fertility drugs and contraceptive agents. In *Towards the prevention of fetal malformation* (ed. J.B. Scrimgeour) pp.89–100. Edinburgh University Press.

—— (1981). The abortus as a predictor of future trisomy 21 pregnancies. In *Trisomy 21 (Down syndrome)* (ed. F.F. de la Cruz and P.S. Gerald) pp.69–76. University Park Press, Baltimore.

—— and Creasy, M.R. (1977). Frequency of chromosomal abnormalities in miscarriages and perinatal deaths. *J. med. Genet.* **14**, 313–15.

—— Polani, P.E., Fraser Roberts, J.A., Spicer, C.C., Elliott, M., and Armstrong, E. (1972). Parental exposure to X-irradiation and Down's syndrome. *Ann. hum. Genet.* **36**, 195–208.

Alexiou, D., Chrysostomidou, O., Vlachos, I., and Deligeorgis, D. (1971). Trisomy 18 with ovarian dysgenesis. *Acta paediat. scand.* **60**, 93–7.

Alfi, O.S., Chang. R., and Azen, S.P. (1980). Evidence for genetic control of nondisjunction in man. *Am. J. hum. Genet.* **32**, 477–83.

Allison, A.C., Hulands, G.H., Nunn, J.F., Kitching, J.A., and MacDonald, A.C. (1970). The effects of inhalational anaesthetics on the microtubular system in *Actinosphaerium nucleofilum. J. Cell Sci.* **7**, 483–99.

Anderson, E.G. (1924). X–rays and the frequency of nondisjunction in *Drosophila. Pap. Mich. Acad. Sci.* **IV**, 523–5.

—— (1929). Studies in a case of high nondisjunction in *Drosophila melanogaster. Z. indukt. Abstamm.-u. VererbLehre* **51**, 397–441.

—— (1931). The constitution of primary exceptions obtained after X-ray treatment of *Drosophila. Genetics* **16**, 386–96.

Auerbach, C. (1976). *Mutation research: problems, results and perspectives.* Chapman and Hall, London.

Austin, C.R. (1967). Chromosome deterioration in ageing eggs of the rabbit. *Nature, Lond.* **213**, 1018–19.

Awa, A.A. (1975). Genetic effects, Cytogenetic study. In *Review of thirty years study of Hiroshima and Nagasaki atomic bomb survivors* (ed. S. Okada, H.B. Hamilton, N. Egami, S. Okajima, W.J. Russell, and K. Takeshita). *J. Radiat. Res.* **16**, Suppl. 75–81.

Bacq, Z.M. and Alexander, P. (1961). *Fundamentals of radiobiology.* Pergamon Press, Oxford.

Baikie, A.G., Court Brown, W.M., Buckton, K.E., and Harnden, D.G. (1961). Two cases of leukaemia and a case of sex-chromosome abnormality in the same sibship. *Lancet* **ii**, 1003–4.

Bajer, A. (1972). Influence of UV microbeam on spindle fine structure and anaphase chromosome movements. In *Chromosomes today* (ed. C.D. Darlington and K. Lewis) Vol. 3, pp.63–9.

—— and Molè-Bajer, J. (1972). Spindle dynamics and chromosome movement. *Int. Rev. Cytol.* Suppl. **3**, 1–271.

Baker, B.S. (1975). Paternal loss (pal): a meiotic mutant in *Drosophila melanogaster* causing loss of paternal chromosomes. *Genetics* **80**, 267—96.

—— Carpenter, A.T.C., Esposito, M.S., Esposito, R.E., and Sandler, L. (1976). The genetic control of meiosis. *A. Rev. Genet.* **10**, 53–134.

—— —— and Ripoll, P. (1978). The utilization during mitotic cell division of loci controlling meiotic recombination and disjunction in *Drosophila*. *Genetics* **90**, 531–78.

Baldwin, M. and Chovnick, A. (1967). Autosomal half-tetrad analysis in *Drosophila melanogaster*. *Genetics* **55**, 277–93.

Barber, H.N. (1940). The suppression of meiosis and the origin of diplochromosomes. *Proc. R. Soc. Lond. B* 170–85.

Barr, M.L., Sergovich, F.R., Carr, D.H., and Shaver, E.L. (1969). The triple-X syndrome. *Can. med. Ass. J.* **101**, 247–58.

Barthelmess, A. (1957). Chemisch induzierte multipolare mitosen. *Protoplasma* **48**, 546–61.

—— (1970). Mutagenic substances in the human environment. In *Chemical mutagenesis in mammals and man* (ed. F. Vogel and G. Röhrborn) pp.69–147. Springer, Berlin.

Bass, H.N., Crandall, B.F., and Sparkes, R.S. (1973). Probable trisomy 22 identified by fluorescent and trypsin-giemsa banding. *Ann. Génét.* **16**, 189–92.

Bateman, A.J. (1960). The induction of dominant lethal mutations in rats and mice by triethylenemelamine (TEM). *Genet. Res.* **1**, 381–92.

—— (1968). Nondisjunction and isochromosomes from irradiation of chromosome 2 in *Drosophila*. In *Effects of radiation on meiotic systems*, pp. 63–70 IAEA.

—— (1969). A storm in a coffee cup. *Mut. Res.* **7**, 475–8.

Battaglia, E. (1964). Cytogenetics of B-chromosomes. *Caryologia* **17**, 246–99.

Beatty, R.A. (1977), F-bodies as Y chromosome markers in mature human sperm heads: a quantitative approach. *Cytogenet. Cell Genet.* **18**, 33–49.

—— Lim, M.-C., and Coulter, V.J. (1975). A quantitative study of the second metaphase in male mice. *Cytogenet. Cell Genet.* **15**, 256–75.

Beechey, C.V. (1973). X-Y chromosome dissociation and sterility in the mouse. *Cytogenet. Cell Genet.* **12**, 60–7.

Bell, A.G. and Corey, P.N. (1974). A sex chromatin and Y-body survey of Toronto newborns. *Can. J. Genet. Cytol.* **16**, 239–50.

Benda, C.E. (1969). *Down's syndrome, mongolism and its management*. Grune and Stratton, New York.

Bennett, M.D. (1982). Nucleotypic basis of the spatial ordering of chromosomes in eukaryotes and the implications of the order for genome evolution and phenotypic variation. In *Genome evolution* (ed. G.A. Dover and R.B. Flavell) pp.239–61. Academic Press, New York.

Berthelsen, J.G., Skakkebaek, N.E., Perbøll, O. and Nielsen, J. (1981). Electron microscopical demonstration of the extra Y chromosome in spermatocytes from human XYY males. In *Development and function of reproductive organs* (ed. A.G. Byskov and H. Peters) Int. Congress Series No. 559. Elsevier/North Holland, Amsterdam.

Bhattacharya, S. (1949). Tests for a possible action of ethylene glycol on the chromosomes of *Drosophila melanogaster*. *Proc. R. Soc. Edinb. B* **63**, 242–8.

Bianchi, N.O. and Contreras, J.R. (1967). The chromosomes of the field mouse, *Akodon azarae* (Cricetidae, Rodentia) with special reference to sex chromosome anomalies. *Cytogenetics* **6**, 306–13.

Bierman, J.M., Siegel, E., French, F.E., and Simonian, K. (1965). Analysis of all pregnancies in a community Kauai pregnancy study. *Am. J. Obstet. Gynec.* **91**, 37–45.

Biesele, J.J. and Jacquez, J.A. (1954). Mitotic effects of certain amino acid analogues in tissue culture. *Ann. NY Acad. Sci.* **58**, 1276–87.

Binkert, F. and Schmid, W. (1977). Pre-implantation embryos of Chinese hamster. I. Incidence of karyotype anomalies in 226 control embryos. *Mut. Res.* **46**, 63–76.

Bobrow, M., Madan, K., and Pearson, P.L. (1972). Staining of some specific regions of human chromosomes, particularly the secondary constriction of No. 9. *Nature, New Biol.* **238**, 122–4.

Bochkov, N.P., Kuleshov, N.P., Chebotarev, A.N., Alekhin, V.I., and Midian, S.A. (1974). Population cytogenetic investigation of newborns in Moscow. *Humangenetik* **22**, 139–52.

Boer, P de and Hoeven, F.A. van der (1980). The use of translocation-derived 'marker bivalents' for studying the origin of meiotic instability in female mice. *Cytogenet. Cell Genet.* **26**, 49–58.

Bond, D.J. (1976). A system for the study of meiotic nondisjunction using *Sordaria brevicollis. Mut. Res.* **37**, 213–20.

—— and McMillan, L. (1979). Meiotic aneuploidy: its origins and induction following chemical treatment in *Sordaria brevicollis. Env. Hlth Persp.* **31**, 67–74.

Böök, J.A. (1945). Cytological studies in *Triton. Hereditas* **31**, 177–220.

Borgaonkar, D.S. and Mules, E. (1970). Comments on patients with sex chromosome aneuploidy: dermatoglyphs, parental ages, Xgᵃ blood group. *J. med. Genet.* **7**, 345–50.

Borisy, G.G. and Taylor, E.W. (1967a). The mechanism of action of colchicine: Binding of colchicine-H^3 to celluar protein. *J. Cell Biol.* **34**, 525–33.

—— —— (1967b). The mechanism of action of colchicine. Colchine binding to sea urchin eggs and the mitotic apparatus. *J. Cell Biol.* **34**, 535–48.

Boué, A. and Thibault, C. (eds.) (1973). *Les accidents chromosomiques de la reproduction.* Institut National de la Santé et de la Racherche Médicale (INSERM), Paris.

Boué, J.G. and Boué, A. (1973a). Anomalies chromosomiques dans les avortements spontanes. In *Les accidents chromosomiques de la reproduction* (ed. A. Boué and C. Thibault) pp. 29–55. Institut National de la Santé et de la Recherche Médicale (INSERM), Paris.

—— —— (1973b). Increased frequency of chromosomal anomalies in abortions after induced ovulation. *Lancet* **i**, 679–80.

—— —— (1973c). Chromosomal analysis of two consecutive abortions in each of 43 women. *Hum. Genet.* **19**, 275–80.

—— —— and Lazar. P. (1975) Retrospective and prospective epidemiological studies of 1500 karyotyped spontaneous human abortions. *Teratology* **12**, 11–26.

—— Deluchat, C., Nicolas, H., and Boué, A. (1981). Prenatal losses of trisomy 21. In *Trisomy 21* (ed. G.R. Burgio, M. Fraccaro, L. Tiepolo, and U. Wolf) pp. 183–93. Springer, Berlin.

Brabander, M. de, Gevens, G., Nuydens, R., Willebroros, R., and De Mey, J. (1981). Taxol induces the assembly of free microtubules in living cells and blocks the organizing capacity of the centromeres and kinetochores. *Proc. natn. Acad. Sci. USA* **78**, 5608–12.

Breeuwsma, A.J. (1968). A case of XXY sex chromosome constitution in an intersex pig. *J. Reprod. Fert.* **16**, 119–20.

Bridges, C.B. (1913). Nondisjunction of the sex-chromosome of *Drosophila. J. exp. Zool.* **15**, 587–606.

—— (1916). Nondisjunction as proof of the chromosome theory of heredity. *Genetics* **1**, 1–52; 107–63.

Brinkley, B.R. and Cartwright, J. (1975). Cold-labile and cold-stable microtubules in the mitotic spindle of mammalian cells. *Am. NY Acad. Sci.* **253**, 428–39.

—— and Rao, N.J. (1973). Nitrous oxide: effects on the mitotic apparatus and chromosome movement in Hela cells. *J. Cell Biol.* **58**, 96–106.

Brook. J.D. (1983). Nondisjunction in mammalian germ cells. Ph.D. thesis. University of Edinburgh.

Brotherston. J. (1978). Implications of antenatal screening for the Health Service. In *Towards the prevention of fetal malformation* (ed. J.B. Scrimgeour) pp. 247–60. Edinburgh University Press.

Bruère, A.N., Marshall, R.B., and Ward. D.P.J. (1969). Testicular hypoplasia and XXY sex chromosome complement in two rams: the ovine counterpart of Klinefelter's syndrome in man. *J. Reprod. Fert.* **19**, 103–8.

Bryan, J. (1972). Definition of three classes of binding sites in isolated microtubule crystals. *Biochemistry* **11**, 2611–16.

Buckton, K.E., O'Riordan, M.L., Ratcliffe, S., Slight, J., Mitchell, M., McBeath, S., Keay, A.J., Barr, D., and Short, M. (1980). A G-band study of chromosomes in liveborn infants. *Ann. hum. Genet.* **43**, 227–39.

Bunge, R.G. and Bradbury, J.T. (1956). Newer concepts of the Klinefelter's syndrome. *J. Urol.* **76**, 758–65.

Burgoyne, P.S. (1978). The role of the sex chromosomes in mammalian germ cell differentiation. *Ann. Biol. anim. Biochem. Biophys.* **18**, 317–25.

—— (1979). Evidence for an association between univalent Y chromosomes and spermatocyte loss in XYY mice and men. *Cytogenet. Cell Genet.* **23**, 84–9.

—— and Biggers, J.D. (1976). The consequences of X-dosage deficiency in the germ line: impaired development *in vitro* of preimplantation embryos from XO mice. *Devl Biol.* **51**, 109–17.

Busby, N. (1971). Segregation following interchange induced by irradiating mature oocytes of *Drosophila melanogaster*. *Mut. Res.* **11**, 391–6.

Buss, M.E. and Henderson, S.A. (1971*a*). Induced bivalent interlocking and the course of meiotic chromosome synapsis. *Nature, New Biol.* **234**, 243–6.

—— —— (1971*b*). The induction of orientational instability and bivalent interlocking at meiosis. *Chromosoma* **35**, 153–83.

Butcher, R.L. and Fugo, N.W. (1967). Overripeness and the mammalian ova II. Delayed ovulation and chromosome anomalies. *Fert. Steril.* **18**, 297–302.

Cacheiro, N.A. and Generoso, W.M. (1975). Cytological studies of sterility in sons of mice treated at spermatogonial or early spermatocyte stages with mutagenic chemicals. *Mouse News Lett.* **53**, 52.

Campbell, D.A. and Fogel, S. (1977). Association of chromosome loss with centromere adjacent mitotic recombination in a yeast disomic haploid. *Genetics* **85**, 573–85.

—— —— and Lusnak, K. (1975). Mitotic chromosome loss in a disomic haploid of *Saccharomyces cerevisiae*. *Genetics* **79**, 383–96.

Cannings, C. and Cannings, M.R. (1968). Mongolism, delayed fertilization and human sexual behaviour. *Nature, Lond.* **218**, 481.

Cappuccinelli, P., Fighetti, M., and Rubino, S. (1979). A mitotic inhibitor for chromosomal studies in slime moulds. *Fems Microb. Lett.* **5**, 25–7.

Carothers, A.D., Collyer, S., de Mey, R., and Frackiewicz, A. (1978). Parental age and birth order in the aetiology of some sex chromosome aneuploidies. *Ann. hum. Genet.* **41**, 277–87.

—— Frackiewicz, A., de Mey, R., Collyer, S., Polani, P.E., Osztovics, M.,

Horvath, K., Papp, Z., May, H.M., and Ferguson-Smith, M.A. (1980). A collaborative study of the aetiology of Turner syndrome. *Ann. hum. Genet.* **43**, 355–68.

Carr, D.H. (1963). Chromosome studies in abortuses and stillborn infants. *Lancet* **ii**, 603–6.

—— (1971). Chromosomes and abortion. In *Advances in human genetics* (ed. H. Harris and K. Hirschhorn) Vol. 2, pp.201–57. Plenum Press, New York.

—— and Gedeon, M. (1977). In *Population cytogenetics. Studies in humans* (ed. E.B. Hook and I.H. Porter) pp.1–9. Academic Press, New York.

—— Haggar, R.A., and Hart, A.G. (1968). Germ cells in the ovaries of XO female infants. *Am. J. clin. Path.* **49**, 521–6.

Carter, C.O. (1958). A life table for mongols with the causes of death. *J. ment. defic. Res.* **2**, 64–74.

Case, M.E. and Giles, N.M. (1964). Allelic recombination in *Neurospora*: tetrad analysis of a three point cross within the pan-2 locus. *Genetics* **49**, 529–40.

Casey, M.D., Segall, L.J., Street, D.R.K., and Blank, C.E. (1966). Sex chromosome abnormalities in two State hospitals for patients requiring special security. *Nature, Lond.* **209**, 641–2.

—— Street, D.R.K., Segall, L.J., and Blank, C.E. (1968). Patients with sex chromatin abnormality in two State hospitals. *Ann. hum. Genet.* **32**, 53–63.

Caspersson, T., Lomakka, G., and Zech, L. (1971). The 24 fluorescence patterns of the human metaphase chromosomes—distinguishing characters and variability. *Hereditas* **67**, 89–102.

—— Hultén, M., Lindsten, J., and Zech, L. (1970). Distinction between extra G-like chromosomes by quinacrine mustard fluorescence analysis. *Expl Cell Res.* **63**, 240–3.

Cattanach, B.M. (1957). The induction of translocations in male mice by triethylenemelamine. *Nature, Lond.* **180**, 1364–5.

—— (1961). XXY mice. *Genet. Res.* **2**, 156–60.

—— (1964). Autosomal trisomy in the mouse. *Cytogenetics* **3**, 159–66.

—— (1967*a*). A test of distributive pairing between two specific non-homologous chromosomes in the mouse. *Cytogenetics* **6**, 67–77.

—— (1967*b*). Induction of paternal sex-chromosome losses and deletions and of autosomal gene mutations by the treatment of mouse post-meiotic germ cells with triethylenemelamine. *Mut. Res.* **4**, 73–82.

—— and Pollard, C.E. (1969). An XYY sex chromosome constitution in the mouse. *Cytogenetics* **8**, 80–6.

Centerwell, W.R. and Benirschke, K. (1973). Male tortoiseshell and calico (T.C.) cats: animal models of sex chromosome mosaics, aneuploids, polyploids and chimaeras. *J. Hered.* **64**, 272–8.

Chandley, A.C. (1966). Studies on oogenesis in *Drosophila melanogaster* with H³-thymidine label. *Expl Cell Res.* **44**, 201–15.

—— (1968). The effect of X-rays on female germ cells of *Drosophila melanogaster*. III A comparison with heat-treatment on crossing over in the X-chromosome. *Mut. Res.* **5**, 93–107.

—— (1979). Chromosomal basis of human infertility. *Br. med. Bull.* **35**, 181–6.

—— and Bateman, A.J. (1962). Timing of spermatogenesis in *Drosophila melanogaster* using H³-thymidine. *Nature, Lond.* **193**, 299–300.

—— and Speed, R.M. (1979). Testing for nondisjunction in the mouse. *Env. Hlth Persp.* **31**, 123–9.

—— Fletcher, J., and Robinson, J.A. (1976). Normal meiosis in two 47,XYY men. *Hum. Genet.* **33**, 231–40.

—— Maclean, N., Edmond, P., Fletcher, J., and Watson, G.S. (1976). Cytogenetics and infertility in man. II Testicular histology and meiosis. *Ann. hum. Genet.* **40**, 165–76.

—— Hargreave, T.B., Fletcher, J.M., Soos, M., Axworthy, D., and Price, W.H. (1980). Trisomy 8: report of a mosaic human male with near-normal phenotype and normal IQ, ascertained through infertility. *Hum. Genet.* **55**, 31–8.

—— Fletcher, J., Rossdale, P.D., Peace, C.K., Ricketts, S.W., McEnery, R.J., Thorne, J.P., Short, R.V., and Allen, W.R. (1975). Chromosome abnormalities as a cause of infertility in mares. In *Equine reproduction. J. Reprod. Fert.* Suppl. **23**, 377–83.

Chrisman, C.L. (1974). Aneuploidy in mouse embryos induced by diethylstilbestrol-diphosphate. *Teratology* **9**, 229–32.

—— and Hinkle, L.L. (1974). Induction of aneuploidy in mouse bone marrow cells with diethylstilbestrol-diphosphate. *Can. J. Genet. Cytol.* **16**, 831–5.

Church, K. and Wimber, D.E. (1969). Meiosis in the grasshopper: chiasma frequency after elevated temperature and X-rays. *Can. J. Genet. Cytol.* **11**, 209–16.

Clark, A.M. and Clark, E.G. (1968). The genetic effects of caffeine in *Drosophila melanogaster*. *Mut. Res.* **6**, 227–34.

—— and Sobels, F.H. (1973). Studies on nondisjunction of the major autosomes in *Drosophila melanogaster*. I Methodology and rate of induction by X-rays for the compound second chromosome. *Mut. Res.* **18**, 47–61.

Clements, J. and Todd, N.K. (1981). Halothane and nondisjunction in *Drosophila*. *Mut. Res.* **91**, 225–8.

Clemon, G.P. and Sisler, H.D. (1969). Formation of a fungitoxic derivative from benlate. *Phytopathology* **59**, 705–6.

Clendenin, T.M. and Benirschke, K. (1963). Chromosome studies on spontaneous abortions. *Lab. Invest.* **12**, 1281–92.

Clough, E., Pyle, R.L., Hare, W.C., Kelly, D.F., and Patterson, D.F. (1970). An XXY sex chromosome constitution in a dog with testicular hypoplasia and congenital heart disease. *Cytogenetics* **9**, 71–7.

Cohen, B.H. and Lilienfeld, A.M. (1970). The epidemiological study of mongolism in Baltimore. In *Down's syndrome (mongolism)* (ed. V. Apgar) *Ann. NY Acad. Sci.* **171**, 320–7.

Cohen, E.N., Bellville, J.W., and Brown, B.W. (1971). Anesthesia, pregnancy and miscarriage. A study of operating room nurses and anesthetists. *Anesthesiology* **35**, 343–7.

Cohen, M.M., Dahan, S., and Shahens, M. (1975). Cytogenetic evaluation of 500 Jerusalem newborn infants. *Israel J. med. Sci.* **11**, 969–77.

Collman, R.D. and Stoller, A. (1962). A survey of mongoloid births in Victoria, Australia. 1942–1957. *Am. J. publ. Hlth* **52**, 813–29.

—— —— (1969). Shift of childbirths to younger mothers and its effect on the incidence of mongolism in Victoria, Australia. 1939–1964. *J. ment. defic. Res.* **13**, 13–19.

Cooke, P. and Curtis, D.J. (1974). General and specific patterns of acrocentric association in parents of mongol children. *Humangenetik* **23**, 279–87.

Cooper, K.W. (1964). Meiotic conjunctive elements not involving chiasmata. *Proc. natn. Acad. Sci. USA* **52**, 1248–55.

Corbett, T.H., Cornell, R.G., Endres, J.L., and Lieding, K. (1974). Birth defects among children of nurse-anaesthetists. *Anesthesiology* **41**, 341–4.

Costoff, A. and Mahesh, V. (1975). Primordial follicles with normal oocytes in the ovaries of postmenopausal women. *J. Am. Geriat. Soc.* **23**, 193–6.

Court Brown, W.M. (1968). Development of knowledge about males with an XYY sex chromosome complement. *J. med. Genet.* **5**, 341–59.

—— Law, P., and Smith, P.G. (1969). Sex chromosome aneuploidy and parental age. *Ann. hum. Genet.* **33**, 1–14.

Cox, D.M. (1973). A quantitative analysis of colcemid induced chromosomal nondisjunction in Chinese hamster cells *in vitro*. *Cytogenet. Cell Genet.* **12**, 165–74.

—— and Puck, T.T. (1969). Chromosomal nondisjunction: the action of colcemid on Chinese hamster cells *in vitro*. *Cytogenetics* **8**, 158–69.

Coyle, M.B. and Pittenger, T.H. (1965). Mitotic recombination in pseudo-wild types of *Neurospora*. *Genetics* **52**, 609–25.

Crackower, S.H.B. (1972). The effects of griseofulvin mitosis in *Aspergillus nidulans*. *Can. J. Microbiol.* **18**, 683–7.

Creasy, M.R. and Crolla, J.A. (1974). Prenatal mortality of trisomy 21 (Down's syndrome). *Lancet* **i**, 473–4.

—— —— and Alberman, E.D. (1976). A cytogenetic study of human spontaneous abortions using banding techniques. *Hum. Genet.* **31**, 177–96.

Crowley, P.H., Gulah, D.K., Hayden, T.L., Lopez, P., and Dyer, R. (1979). A chiasma-hormonal hypothesis relating Down's syndrome and maternal age. *Nature, Lond.* **280**, 417–18.

Cunha, F. Da. (1970). Mitotic mapping of *Schizosaccharomyces pombe*. *Genet. Res.* **16**, 127–44.

Danford, N. Meiotic aneuploidy in mice: detection with X-linked coat colour alleles. (Abstr.) *Mut. Res.* in press.

Davidse, L.D. and Flach, W. (1977). The mechanism of resistance to the anti tubulin methyl-benzimidazol-2-yl carbamate in the fungus *Aspergillus nidulans*. *J. Cell Biol.* **72**, 174–93.

Davis, B.K. (1971). Genetic analysis of a meiotic mutant resulting in precocious sister-centromere separation in *Drosophila melanogaster*. *Molec. gen. Genet.* **113**, 251–72.

Davison, E.V., Roberts, D.F., and Callow, M.H. (1981). Satellite association in Down's syndrome. *Hum. Genet.* **56**, 309–13.

Dävring, L. and Sunner, M. (1973). Female meiosis and embryonic mitosis in *Drosophila melanogaster*. I Meiosis and fertilization. *Hereditas* **73**, 51–64.

—— —— (1976). Early prophase in female meiosis of *Drosophila melanogaster*, Further studies. *Hereditas* **82**, 129–31.

—— —— (1977). Late prophase and first metaphase in female meiosis of *Drosophila melanogaster*. *Hereditas* **85**, 25–32.

Day, A.W. and Jones, J.K. (1971). *p*-Fluorophenylalanine-induced mitotic haploidization in *Ustilago violaceae*. *Genet. Res.* **18**, 299–309.

Delhanty, J.D.A., Ellis, J.R., and Rowley, P.T. (1961). Triploid cells in a human embryo. *Lancet* **i**, 1286.

Demerec, M. and Farrow, J.G. (1930*a*). Nondisjunction of the X-chromosome in *Drosophila virilis*. *Proc. natn. Acad. Sci. USA* **16**, 707–10.

—— —— (1930*b*). Relation between the X-ray dosage and the frequency of primary nondisjunction of X-chromosomes in *Drosophila virilis*. *Proc. natn. Acad. Sci. USA* **16**, 711–14.

Donahue, R.P. (1972). Cytogenetic analysis of the first cleavage division in mouse embryos. *Proc. natn. Acad. Sci. USA* **69**, 74–7.

Dover, G.A. and Riley, R. (1973). Effect of spindle inhibitors applied before meiosis on meiotic chromosome pairing. *J. Cell Sci.* **12**, 143–61.

Driscoll, C.J. and Darvey, L. (1970). Chromosome pairing: effect of colchicine on an isochromosome. *Science, NY* **169**, 290–1.

—— —— and Barber, H.N. (1967). Effect of colchicine on meiosis of hexaploid wheat. *Nature, Lond.* **216**, 687–8.

Dumon, J.E. and Leroy, J.G. (1980). A.I.D. and recurrent nondisjunction. *Clin. Genet.* **17**, 62.

Dutrillaux, B. and Lejeune, J. (1970). Etude de la decendance des individus porteurs d'une translocation t(DqDq). *Ann. Génét.* **13**, 11–19.

Edwards, J.H. (1961). Seasonal incidence of congenital disease in Birmingham. *Ann. hum. Genet.* **25**, 89–93.

—— Harnden, D.G., Cameron, A.H., Crosse, V.M., and Wolff, O.H. (1960). A new trisomic syndrome. *Lancet* **i**, 787–90.

Edwards, R.G. (1954). Colchicine-induced heteroploidy in early mouse embryos. *Nature, Lond.* **174**, 276–7.

—— (1958). Colchicine-induced heteroploidy in the mouse. I The induction of triploidy by treatment of the gametes. *J. exp. Zool.* **137**, 317–47.

—— (1965). Maturation *in vitro* of mouse, sheep, cow, pig, rhesus monkey and human ovarian oocytes. *Nature, Lond.* **208**, 349–51.

Epstein, C.J. (1981). Animal models for autosomal trisomy. In *Trisomy 21 (Down syndrome)* (ed. F.F. de la Cruz and P.S. Gerald) pp.263–74. University Park Press, Baltimore.

Erhardt, C.L. (1963). Pregnancy losses in New York City, 1960. *Am. J. publ. Hlth* **53**, 1337–52.

Erickson, J.D. (1978). Down syndrome, paternal age, maternal age and birth order. *Ann. hum. Genet.* **41**, 289–98.

Evans, E.P., Burtenshaw, M.D., and Ford, C.E. (1972). Chromosomes of mouse embryos and newborn young: preparations from membranes and tail tips. *Stain Technol.* **47**, 229–34.

—— Ford, C.E., and Searle, A.G. (1969). A39,X/41,XYY mosaic mouse. *Cytogenetics* **8**, 87–96.

—— —— Chaganti, R.S.K., Blank, C.E., and Hunter, H. (1970). XY spermatocytes in an XYY male. *Lancet* **i**, 719–20.

Evans, H.J. (1967). The nucleolus, virus infection and trisomy in man. *Nature, Lond.* **214**, 361–3.

Fabricant, J.D. and Schneider, E.L. (1978). Studies on the genetic and immunologic components of the maternal age effect. *Devl Biol.* **66**, 337–43.

Fankhauser, G. (1934). Cytological studies on egg fragments of the salamander *Triton.* V Chromosome number and chromosome individuality in the cleavage mitoses of merogonic fragments. *J. exp. Zool.* **68**, 1–57.

Fechheimer, N.S. and Beatty, R.A. (1974). Chromosomal abnormalities and sex ratio in rabbit blastocysts. *J. Reprod. Fert.* **37**, 331–41.

Feingold, M. and Atkins, L. (1973). A case of trisomy 9. *J. med. Genet.* **10**, 184–7.

Ferguson, J., Hawkins, S.W., and Doxey D. (1950). C-mitotic action of single gases. *Nature, Lond.* **165**, 1021–2.

Ferguson-Smith, M.A. (1959). The prepubertal testicular lesion in chromatin-positive Klinefelter's syndrome (primary micro-orchidism) as seen in mentally handicapped children. *Lancet* **i**, 219–22.

—— (1965). Karyotype-phenotype correlations in gonadal dysgenesis and their bearing on the pathogenesis of malformations. *J. med. Genet.* **2**, 142–55.

—— and Handmaker, S.D. (1961). Observations on the satellited human chromosomes. *Lancet* **i**, 638.

—— and Munro, I.B. (1958). Spermatogenesis in the presence of female nuclear sex. *Scott. med. J.* **3**, 39–42.

—— Mack, W.S., Ellis, P.M., Dickson, M., Sanger, R., and Race. R.R. (1964). Parental age and the source of the X-chromosome in XXY Klinefelter's syndrome. *Lancet* **i**, 46.

Ferreira, N.R. and Buoniconti, A. (1968). Trisomy after colchicine therapy. *Lancet* **ii**, 1304.

Fetner, R.H. and Porter, E.D. (1965). Multipolar mitosis in the KB (eagle) human cell line and its increased frequency as a function of 250 Kv X-irradiation. *Expl Cell Res.* **37**, 429–39.

Fialkow, P.J., Uchida, I.A., Hecht, F., and Motulsky, A.G. (1965). Increased frequency of thyroid auto-antibodies in mothers of patients with Down's syndrome. *Lancet* **ii**, 868–70.

Finch, R.A., Böök, J.A., Finley, W.H., Finley, S.C., and Tucker, G.C. (1966). Meiosis in trisomic Down's syndrome. *Ala. J. med. Sci.* **3**, 117–20.

Fiskesjo, G. (1970). The effect of two organic mercury compounds on human leukocytes *in vitro. Hereditas* **64**, 142–6.

Fitzgerald, P.H., Pickering, A.F., Mercer, J.M., and Miethke, P.M. (1975). Premature centromere division: a mechanism of nondisjunction causing X-chromosome aneuploidy in somatic cells of man. *Ann. hum. Genet.* **38**, 417–28.

Forbes, C. (1960). Non-random assortment in primary nondisjunction in *Drosophila melanogaster. Proc. natn. Acad. Sci. USA* **46**, 222–5.

Ford, C.E. (1975). The time in development at which gross genome imbalance is expressed. In *The early development of mammals* (ed. M. Balls and A.E. Wild) pp.285–304. (Brit. Soc. for Dev. Biol. Symp. 2.) Cambridge University Press.

—— (1981). Nondisjunction. In *Trisomy 21* (ed. G.R. Burgio, M. Fraccaro, L. Tiepolo, and U. Wolf) pp.103–43. Springer, Berlin.

—— and Evans, E.P. (1973). Non-expression of genome imbalance in haplophase and early diplophase of the mouse and incidence of karyotypic abnormality in post-implantation embryos. In *Les accidents chromosomiques de la reproduction* (ed. A. Boué and C. Thibault) pp. 271–85. Institut National de la Santé et de la Recherche Médicale (INSERM), Paris.

—— Searle, A.G., Evans, E.P., and West. B.J. (1969). Differential transmission of translocations induced in spermatogonia of mice by irradiation. *Cytogenetics* **8**, 447–70.

Ford, J.H. (1971). Segregation of univalents on mini spindles. *Nature, Lond.* **229**, 570–1.

—— and Lester, P. (1982). Factors affecting the displacement of human chromosomes from the metaphase plate. *Cytogenet. cell Genet.* **33**, 327–32.

Forer, A. (1965). Local reduction of spindle fiber birefringence in living *Nephrotoma suturalis* spermatocytes induced by ultraviolet microbeam irradiation. *J. Cell Biol.* **25**, No.1 Part 2, 95–117.

—— (1966). Characterization of the mitotic traction system and evidence that birefringent spindle fibres neither produce nor transmit force for chromosome movement. *Chromosoma* **19**, 44–98.

—— (1969). Chromosome movement during cell division. In *Handbook of molecular cytology* (ed. A. Lima-de-Faria) pp.554–601. North Holland, Amsterdam.

Foss, G.L. and Lewis, F.J.W. (1971). A study of four cases with Klinefelter's syndrome, showing motile spermatozoa in their ejaculates. *J. Reprod. Fert.* **25**, 401–8.

Foureman, P.A. (1979). A translocation X:Y system for detecting meiotic nondisjunction and chromosome breakage in males of *Drosophila melanogaster*. *Envir. Hlth Perspect.* **31**, 53–8.

Fraser, J. and Mitchell, A. (1876). Kalmuck idiocy: report of a case with autopsy with notes on 62 cases by A. Mitchell. *J. ment. Sci.* **22**, 169–79.

Fraser, L.R. and Maudlin, I. (1979). Analysis of aneuploidy in first cleavage mouse embryos fertilized *in vitro* and *in vivo*. *Envir. Hlth Perspect.* **31**, 141–50.

Froland, A. (1967). Seasonal dependence in birth of patients with Klinefelter's syndrome. *Lancet* **ii**, 771.

Fugo, N.W. and Butcher, R.L. (1971). Effects of prolonged estrous cycles on reproduction in aged rats. *Fert. Steril.* **22**, 98–101.

Galloway, S.M. and Buckton, K.E. (1978). Aneuploidy and ageing: chromosome studies on a random sample of the population using G-banding. *Cytogenet. Cell Genet.* **20**, 78–95.

Gelbart, W. (1974). A new mutant controlling mitotic chromosome disjunction in *Drosophila melanogaster*. *Genetics* **76**, 51–63.

Gelei, G. von and Csik, L. (1939). A colchicin hatása a *Drosophila melanogasterre*. *Arb. ung. bio. ForschInst.* **xi**, 50–63.

Generoso, W.M., Cain, K.T., and Huff, S.W. (1973). Chemical induction of sex-chromosome loss in female mice. Oak Ridge Nat. Lab. Biol. Div. Ann. Progr. Rep. June 30, ORNL–4915, 111–12.

Geraedts, J. and Pearson, P. (1973). Specific staining of the human No. 1 chromosome in spermatozoa. *Humangenetik* **20**, 171–3.

German, J. (1968). Mongolism, delayed fertilization and human sexual behaviour. *Nature, Lond.* **217**, 516–18.

Gershenson, S. (1935). The mechanism of nondisjunction in the ClB stock of *Drosophila melanogaster*. *J. Genet.* **30**, 115–25.

Giménez-Martín, G. and López-Sáez, V.F. (1974). Mutagénesis por acción del j-hexaclorociclohexano. *Phyton* **16**, 45–55.

Goad, W.B., Robinson, A., and Puck. T.T. (1976). Incidence of aneuploidy in a human population. *Am. J. Hum. Genet.* **28**, 62–8.

Golbus, M.S. (1981). The influence of strain, maternal age and method of maturation on mouse oocyte aneuploidy. *Cytogenet. Cell Genet.* **31**, 84–90.

Goldstein, L.S.B. (1980). Mechanisms of chromosome orientation revealed by two meiotic mutants in *Drosophila melanogaster*. *Chromosoma* **78**, 79–111.

Gonzales, M.A. (1967). Formacion y desarollo de celulas binucleadas: bimitosis. *Genet. iber.* **19**, 1–98.

Goodlin, R.C. (1965). Nondisjunction and maternal age in the mouse. *J. Reprod. Fert.* **9**, 355–6.

Goodpasture, C. and Bloom, S.E. (1975). Visualization of nucleolar organizer regions in mammalian chromosomes using silver staining. *Chromosoma* **53**, 37–50.

Gordon, R.R. and O'Neill, E.M. (1969). Turner's infantile phenotype. *Br. med. J.* **i**, 483–5.

Gosden, R.G. (1973). Chromosomal anomalies of preimplantation mouse embryos in relation to maternal age. *J. Reprod. Fert.* **35**, 351–4.

—— and Walters, D.E. (1974). Effects of low-dose X-irradiation on chromosomal nondisjunction in aged mice. *Nature, Lond.* **248**, 54–5.

Grant, C.J., Powell, J.N., and Radford, S.G. (1977). The induction of chromosomal abnormalities by inhalation anaesthetics. *Mut. Res.* **46**, 177–84.

Greenberg, R.C. (1964). Some factors in the epidemiology of mongolism. *Proc. Int. Copenhagen Cong. Sci. Study ment. Retard.* Vol. 1, p.200.

Grell, E.H. (1963). Distributive pairing of compound chromosomes in females of *Drosophila melanogaster. Genetics* **48**, 1217–29.

—— (1970). Distributive pairing: mechanism for segregation of compound autosomal chromosomes in oocytes of *Drosophila melanogaster. Genetics* **65**, 65–74.

Grell, R.F. (1959). Non-random assortment of non-homologous chromosomes in *Drosophila melanogaster. Genetics* **44**, 421–35.

—— (1962*a*). A new model for secondary nondisjunction: the role of distributive pairing. *Genetics* **47**, 1737–54.

—— (1962*b*). A new hypothesis on the nature and sequence of meiotic events in the female *Drosophila melanogaster. Proc. natn. Acad. Sci. USA* **48**, 165–72.

—— (1967). Pairing at the chromosomal level. *J. cell physiol.* **70**, Suppl. 1, 119–45.

—— (1971*a*). Distributive pairing in man? *Ann. Génét.* **14**, 165–71.

—— (1971*b*). Induction of sex chromosome nondisjunction by elevated temperature. *Mut. Res.* **11**, 347–9.

—— (1976). Distributive pairing. In *Genetics and biology of* Drosophila (ed. M. Ashburner and E. Novitski) Vol. 1a, pp. 436–83. Academic Press, New York.

—— and Day, J.W. (1970). Chromosome pairing in the oogonial cells of *Drosophila melanogaster. Chromosoma* **31**, 434–45.

——and Valencia, J.I. (1964). Distributive pairing and aneuploidy in man. *Science, NY* **145**, 66–7.

—— Munoz, E.R., and Kirschenbaum, W.F. (1966). Radiation induced nondisjunction and loss of chromosomes in *Drosophila melanogaster* females. I The effect of chromosome size. *Mut. Res.* **3**, 494–502.

Griffen, A.B. and Bunker, M.C. (1964). Three cases of trisomy in the mouse. *Proc. natn. Acad. Sci. USA* **52**, 1194–8.

—— —— (1967). Four further cases of autosomal primary trisomy in the mouse. *Proc. natn. Acad. Sci. USA* **58**, 1446–52.

Griffiths, A.J.F. (1979). *Neurospora* prototroph selection system for studying aneuploid production. *Envir. Hlth Perspect.* **31**, 75–80.

—— and Delange, A.M. (1977). *p*-Fluorophenylalanine increases meiotic nondisjunction in a *Neurospora* test system. *Mut. Res.* **46**, 345–54.

Grisham, L.M., Wilson, L., and Bensch, K.G. (1973). Antimitotic action of griseofulvin does not involve disruption of microtubules. *Nature, New Biol.* **244**, 294–6.

Gropp, A. and Epstein, C.J. (1982). Value of an animal model for trisomy. In *Progress in clinical and biological research*. Proc. 6th Int. Congress of Human Genetics, Jerusalem, 13–18 Sept. 1981. Liss, New York.

—— Giers, D., and Kolbus, U. (1974). Trisomy in the fetal backcross progeny of male and female metacentric heterozygotes of the mouse I. *Cytogenet. Cell Genet.* **13**, 511–35.

—— Kolbus, U., and Giers D. (1975). Systematic approach to the study of trisomy in the mouse II. *Cytogenet. Cell Genet.* **14**, 42–62.

Grouchy, J. de (1970). 21 p-maternal en double exemplaire chez un trisomique 21. *Ann. Génét.* **13**, 52–5

Guillemin, C. (1980). Meiosis in four trisomic and one double trisomic males of the newt *Pleuorodeles waltlii. Chromosoma* **77**, 145–55.

Gull, K. and Trinci, A.P.J. (1973). Griseofulvin inhibits fungal mitosis. *Nature, Lond.* **244**, 292–4.

—— —— (1974). Ultrastructural effects of griseofulvin on the myxomycete *Physarum polycephalum*. Inhibition of mitosis and the production of microtubule crystals. *Protoplasma* **81**, 37–48.

Gustavson, K.H., Hitrec, V., and Santesson, B. (1972). Three non-mongoloid patients of similar phenotype with an extra G-like chromosome. *Clin. Genet.* **3**, 135–46.

Haber, J.E. (1974). Bisexual mating behaviour in a diploid of *Saccharomyces cerevisiae*: evidence for a genetically controlled non-random loss during vegetative growth. *Genetics* **78**, 843–58.

—— Peloquin, J.G., Halvorson, H.O., and Borisy, G.G. (1972). Colcemid inhibition of cell growth and the characterization of a colcemid binding activity in *Saccharomyces cerevisiae*. *J. cell Biol.* **55**, 355–67.

Hagard, S. and Carter, F.A. (1976). Preventing the birth of infants with Down's syndrome: a cost benefit analysis. *Br.med.J.* **i**, 753–6.

Hall, J.C. (1970). Nonindependence of primary nondisjunction for the sex and fourth chromosomes in *Drosophila melanogaster*. *Drosophila Inform. Serv.* **45**, 160.

Hamerton, J.L. (1971a). General cytogenetics. In *Human cytogenetics*, Vol. 1. Academic Press, New York.

—— (1971b). Clinical cytogenetics. In *Human cytogenetics*, Vol. II. Academic Press, New York.

—— Gianelli, F., and Polani, P.E. (1965). Cytogenetics of Down's syndrome (mongolism). 1. Data on a consecutive series of patients referred for genetic counselling and diagnosis. *Cytogenetics* **4**, 171–89.

—— Canning, N., Ray, M., and Smith, S. (1975). A cytogenetic survey of 14,069 newborn infants. *Clin. Genet.* **8**, 223–43.

Hansmann, I. (1974). Chromosome aberrations in metaphase II oocytes. Stage sensitivity in the mouse oogenesis to amethopterin and cyclophosphamide. *Mut. Res.* **22**, 175–91.

—— and El-Nahass, E. (1979). Incidence of nondisjunction in mouse oocytes. *Cytogenet. Cell Genet.* **24**, 115–21.

—— and Probeck, H.D. (1979). Chromosomal imbalance in ovulated oocytes from Syrian hamster. (*Mesocricetus auratus*) and Chinese hamster (*Cricetulus griseus*). *Cytogenet. Cell Genet.* **23**, 70–6.

Hansson, A. (1979). Satellite association in human metaphases. A comparative study of normal individuals, patients with Down's syndrome and their parents. *Hereditas* **90**, 59–83.

—— and Mikkelsen, M. (1978). The origin of the extra chromosome 21 in Down syndrome. *Cytogenet. Cell Genet.* **20**, 194–203.

Hardy, R.W. (1975). The influence of chromosome content on the size and shape of sperm heads in *Drosophila melanogaster* and the demonstration of chromosome loss during spermiogenesis. *Genetics* **79**, 231–64.

Harlap, S., Shiono, P.H., and Ramcharan, S. (1980). Spontaneous foetal losses in women using different contraceptives around the time of conception. *Int. J. Epidemiol.* **9**, 49–56.

—— —— Pellegrin, F., Golbus, M., Bachman, R., Mann, J., Schmidt, L., and Lewis, J.P. (1979). Chromosome abnormalities in oral contraceptive breakthrough pregnancies. *Lancet* **i**, 1342–3.

Harrison, C.M.H., Page, B.M., and Keir, H.M. (1976). Mescaline as a mitotic spindle inhibitor. *Nature, Lond.* **260**, 138–9.

Hartl, D.L., Miraizumi, Y., and Crow, J.F. (1967). Evidence for sperm dysfunction as the mechanism of segregation distortion in *Drosophila melanogaster*. *Proc. natn. Acad. Sci. USA* **58**, 2240–5.

Hassold, T.J. (1980). A cytogenic study of repeated spontaneous abortions. *Am. J. hum. Genet.* **32**, 723–30.

—— and Matsuyama, A. (1979). Origin of trisomies in human spontaneous abortions. *Hum. Genet.* **46**, 285–94.

—— Jacobs, P., Kline, J., Stein, Z., and Warburton, D. (1980*a*). Effect of maternal age on autosomal trisomies. *Ann. hum. Genet.* **44**, 29–36.

—— Matsuyama, A., Newlands, I.M., Matsura, J.S., Jacobs, P.A., Manuel, B., and Tsuei, J. (1978). A cytogenetic study of spontaneous abortions in Hawaii. *Ann. hum. Genet.* **41**, 443–54.

—— Chen, N., Funkhouser, J., Jooss, T., Manuel, B., Matsuura, J., Matsuyama, A., Wilson, C., Yamane, J.A., and Jacobs, P.A. (1980*b*). A cytogenetic study of 1,000 spontaneous abortions. *Ann. hum. Genet.* **44**, 151–78.

Hastie, A.C. (1970). Benlate induced instability of *Aspergillus* diploids. *Nature, Lond.* **226**, 771.

Heath, I.B. (1975). The effect of antimicrotubule agents on the growth and ultrastructure of the fungus *Saprolegnia ferax* and their ineffectiveness in disrupting hyphal microtubules. *Protoplasmia* **85**, 147–76.

Hecht, F. (1982). Unexpected encounters in cytogenetics: repeated abortions and parental sex chromosome mosaicism may indicate risk of nondisjunction. *Am. J. hum. Genet.* **34**, 514–16.

—— Bryant, J.S., Gruber, D., and Townes, P.L. (1964). The nonrandomness of chromosomal abnormalities. Association of trisomy 18 and Down's syndrome. *New Engl. J. Med.* **271**, 1081.

Held, L.I. (1982). Polyploidy and aneuploidy induced by colcemid in *Drosophila melanogaster*. *Mut. Res.* **94**, 87–101.

Henderson, S.A. (1962). Temperature and chiasma formation in *Schistocerca gregaria*. II Cytological effects of 40°C and the mechanisms of heat-induced univalence. *Chromosoma, Berl.* **13**, 437–63.

—— and Edwards, R.G. (1968). Chiasma frequency and maternal age in mammals. *Nature, Lond.* **218**, 22–8.

—— Nicklas, B., and Koch, C.A. (1970). Temperature-induced orientation instability during meiosis: an experimental analysis. *J. Cell Sci.* **6**, 325–50.

Hendricks, C.H. (1955). Congenital malformations, analysis of the 1953 Ohio records. *Obstet. Gynec.* **6**, 592–8.

Henshaw, P.S. (1940). Further studies on the action of roentgen rays on the gametes of *Arbacia punctulata* VI. Production of multipolar cleavage in the eggs by exposure of the gametes to roentgen rays. *Am. J. Roentgenol radium Ther. nucl. Med.* **43**, 923–33.

Herich, R. (1969). Effect of zinc on mitosis. *Naturwissenschaften* **56**, 286–7.

Hertig, A.T., Rock, J., and Adams, E.C. (1956). A description of 34 human ova within the first 17 days of development. *Am. J. Anat.* **98**, 435–93.

Hertwig, O. (1898). Ueber den Einfluss der Temperatur auf die Entwicklung von *Rana fusca* und *Rana esculenta*. *Arch. Mikroscop. Anat.* **51**, 319–81.

Hess, F.D. and Bayer, D.E. (1977). Binding of the herbicide trifluralin to *Chlamydomonas* flagella tubulin. *J. Cell Sci.* **24**, 351–60.

Hildreth, P.E. and Ulrichs, P.C. (1969). A temperature effect on nondisjunction of the X-chromosome among eggs from aged *Drosophila* females. *Genetica*, **40**, 191–7.

Hoebeke, J., Van Nijen, G., and De Brabander, M. (1976). Interaction of oncodazole (R17934), a new anti-tumoral drug with rat brain tubulin. *Biochem. Biophys. Res. Commun.* **69**, 319–24.

Hoefnagel, D. (1969). Trisomy after colchicine therapy. *Lancet* i, 1160.

Højager, B., Peters, H., Byskov, A.G., and Faber, M. (1978). Follicular development in ovaries of children with Down's syndrome. *Acta paediat. scand.* **67**, 637–43.

Holden, H.E., Florio, J.R., Wahrenburg, M.G., and Ray, V.A. (1976). Drug induced centromere stickiness in mouse chromosomes. *Mam. Chrom. Newsl.* **17**, 31.

Holm, D.G. and Chovnick, A. (1975). Compound autosomes in *Drosophila melanogaster*: the meiotic behaviour of compound thirds. *Genetics* **81**, 293–311.

Hook, E.B. and Porter, I.H. (1977). Comments on racial differences in frequency of chromosome abnormalities. Putative clustering of Down's syndrome and radiation studies. In *Population cytogenetics* (ed. E.B. Hook and I.H. Porter). Academic Press, New York.

Hotta, Y., Chandley, A.C., and Stern, H. (1977). Meiotic crossing-over in lily and mouse. *Nature, Lond.* **269**, 240–2.

Hsu, T.C., Cooper, J.E.K., Mace, M.L., and Brinkley, B.R. (1971). Arrangement of centromeres in mouse cells. *Chromosoma* **34**, 73–87.

Hughes-Schrader, S. (1948). Expulsion of the sex chromosome from the spindle in spermatocytes of a mantid. *Chromosoma* **3**, 257–70.

Hultén, M., and Lindsten, J. (1970). The behaviour of structural aberrations at male meiosis; information from man. In *Human population cytogenetics* (ed. P.A. Jacobs, W.H. Price, and P. Law) Pfizer Medical Monographs, Vol. 5, pp.24–61. Edinburgh University Press.

—— and Pearson, P.L. (1971). Fluorescent evidence for spermatocytes with two Y chromosomes in an XYY male. *Ann. hum. Genet.* **34**, 273–6.

Inoué, S. (1952). Effect of temperature on the birefringence of the mitotic spindle. *Biol. Bull.* **103**, 316.

—— (1964). Organization and function of the mitotic spindle. In *Primitive motile systems in cell biology* (ed. R.D. Allen and N. Kamiya) pp.549–98. Academic Press, New York.

Izutsu, K. (1961*a*). Effects of ultravoilet microbeam irradiation upon division in grasshopper spermatocytes I. Results of irradiation during prophase and prometaphase. *Mie med. J.* **II**, 199–211.

—— (1961*b*). Effects of ultraviolet microbeam irradiation upon division in grasshopper spermatocytes II. Results of irradiation during metaphase and anaphase I. *Mie med. J.* **II**, 213–32.

Jacobs, P.A. (1972). Chromosome mutations: frequency at birth in humans. *Humangenetik* **16**, 137–40.

—— (1979). The incidence and etiology of sex chromosome abnormalities in man. *Birth Defects: Original Article Series*, Vol. XV, No.1, pp.3–14. The National Foundation.

—— and Hassold, T.J. (1980). The origin of chromosome abnormalities in spontaneous abortions. In *Human embryonic and fetal death* (ed. I.H. Porter and E.B. Hook) pp.289–98. Academic Press, New York.

—— and Morton, N.M. (1977). Origin of human trisomics and polyploids. *Hum. Hered.* **27**, 59–72.

—— Brittain, R.P., and McClermont, W.F. (1965). Aggressive behaviour, mental subnormality and the XYY male. *Nature, Lond.* **208**, 1351.

—— Brunten, M., Court Brown, W.M., Doll, R., and Goldstein, H. (1963). Change of human chromosome count distributions with age: evidence for a sex difference. *Nature, Lond.* **197**, 1080–1.

—— Melville, M., Ratcliffe, S., Keay, A.J., and Syme, J. (1974). A cytogenetic survey of 11,680 newborn infants. *Ann. hum. Genet.* **37**, 359–76.

—— Price, W.H., Court Brown, W.M., Brittain, R.P., and Whatmore, P.B.

(1968). Chromosome studies on men in a maximum security hospital. *Ann. hum. Genet.* **31**, 339–47.

—— Baikie, A.G., Court Brown, W.M., Macgregor, T.N., Maclean, N., and Harnden, D.G. (1959). Evidence for the existence of the human 'super female'. *Lancet* **ii**, 423–5.

Jagiello, G. and Fang, J.S. (1979). Analyses of diplotene chiasma frequencies in mouse oocytes and spermatocytes in relation to ageing and sexual dimorphism. *Cytogenet. Cell Genet.* **23**, 53–60.

—— and Lin, J.S. (1973). An assessment of the effects of mercury on the meiosis of mouse ova. *Mut. Res.* **17**, 93–9.

—— —— (1982). Oral contraceptive compounds and mammalian oocyte meiosis. *Am J. Obstet. Gynec.* **129**, 390–406.

—— Karnicki, J., and Ryan, R.J. (1968). Superovulation with pituatory gonadotrophins. Methods for obtaining meiotic metaphase figures in human ova. *Lancet* **i**, 178–80.

James, W.H. (1978). Down syndrome and parental coital rate. *Lancet* **ii**, 895.

Jirásek, J.E. (1977). Morphogenesis of the genital system in the human. In *Birth Defects: Original Article Series*, Vol. 13 (ed. R.J. Blandau and D. Bergsma) pp.13–59. Liss, New York.

Johnston, A.W., Ferguson-Smith, M.A., Handmaker, S.D., Jones, H.W., and Jones, G.S. (1961). The triple-X syndrome. *Br. med. J.* **ii**, 1046–52.

Jones, D.C. and Lowry, R.B. (1975). Falling maternal age and incidence of Down syndrome. *Lancet* **i**, 753–4.

Jones, E.C. and Krohn, P.L. (1961). The relationships between age, numbers of oocytes and fertility in virgin and multiparous mice. *J. Endocr.* **21**, 469–95.

Jones, K.W., and Corneo, G. (1971). Location of satellite and homogeneous DNA sequences on human chromosomes. *Nature, New Biol.* **233**, 268–71.

Jongbloet, P.H. (1971). Month of birth and gametopathy. *Clin. Genet.* **2**, 315–30.

—— Frants, R.R., and Hamers, A.J. (1981). Parental αl-antitrypsin (Pl) types and meiotic nondisjunction in the aetiology of Down syndrome. *Clin. Genet.* **20**, 304–9.

—— Poestkoke, A.J.M., Hamers, A.J.H., and van Erkelens-Zwets, J.H.J. (1978). Down syndrome and religious groups. *Lancet* **ii**, 1310.

Juberg, R.C. and Jones, B. (1970). The Christchurch chromosome (Gp-). Mongolism, erythroleukaemia and an inherited Gp- chromosome (Christchurch). *New Engl. J. Med.* **282**, 292–7.

Kajii, T. and Ohama, K. (1979). Inverse maternal age effect in monosomy X. *Hum. Genet.* **51**, 147–51.

—— —— and Mikamo, K. (1978) Anatomic and chromosomal anomalies in 944 induced abortuses. *Hum. Genet.* **43**, 247–58.

—— —— Niikawa, N., Ferrier, A., and Avirachan, S. (1973). Banding analysis of abnormal karyotypes in spontaneous abortion. *Am. J. hum. Genet.* **25**, 539–47.

Kamiguchi, Y., Funaki, K. and Mikamo, K. (1976). A new technique for chromosome study of murine oocytes. *Proc. Jap. Acad.* **52**, 316–19.

—— —— —— (1979). Chromosomal anomalies caused by preovulatory overripeness of the primary oocyte. *Proc. Jap. Acad.* **55**, 398–402.

Kapp, R.W. (1979). Detection of aneuploidy in human sperm. *Envir. Hlth Perspect.* **31**, 27–31.

Kappas, A. (1978). On the mechanism of induced somatic recombination by certain fungicides in *Aspergillus nidulans*. *Mut. Res.* **51**, 189–97.

—— Georgopulos, S.G., and Hastie. A.C. (1974). On the genetic activity of benzimidazole and thiophanate fungicides on diploid *Aspergillus nidulans. Mut. Res.* **26**, 17–27.

Karp, L.E. and Smith, W.D. (1975). Experimental production of aneuploidy in mouse oocytes. *Gynec. Invest.* **6**, 337–41.

Kato, H. and Yoshida, T.H. (1970). Nondisjunction of chromosomes in a synchronized cell population initiated by reversal of Colcemid inhibition. *Expl Cell Res.* **60**, 459–64.

—— —— (1971). Isolation of aneusomic clones from Chinese hamster cell line following induction of nondisjunction. *Cytogenetics* **10**, 392–403.

Kaufman, M.H. (1973). Analysis of the first cleavage division to determine the sex ratio and incidence of chromosome anomalies of conception in the mouse. *J. Reprod. Fert.* **35**, 67–72.

—— (1977). Effect of anaesthetic agents on eggs and embryos. In *Development of mammals*, Vol. 1 (ed. M.H. Johnson) pp.137–63. Elsevier/North Holland, Amsterdam.

Kihlman, B. and Levan, A. (1949). The cytological effect of caffeine. *Hereditas* **35**, 109–11.

King, C.R., Magenis, E., and Bennett, S. (1978). Pregnancy and the Turner syndrome. *Obstet. Gynec.* **52**, 617–24.

King, R.C. (1957). Oogenesis in adult *Drosophila melanogaster* II. Stage distribution as a function of age. *Growth* **21**, 95–102.

—— Rubinson, A.C., and Smith, R.F. (1956). Oogenesis in adult *Drosophila melanogaster. Growth* **20**, 121–57.

Kjessler, B. and de la Chapelle, A. (1971). Meiosis and spermatogenesis in two post-pubertal males with Down's syndrome. 47,XY,G+. *Clin. Genet.* **2**, 50–7.

Klasterska, I. and Ramel, C.C. (1978). The effects of methyl mercury hydroxide on meiotic chromosomes of the grasshopper *Stethophyma grossum. Hereditas* **88**, 255–62.

Klinefelter, H.F. Jr, Reifenstein, E.C. Jr, and Albright, F. (1942). Syndrome characterized by gynaecomastia aspermatogenesis without a-Leydigism, and increased excretion of follicle stimulating hormone. *J. clin. Endocr.* **2**, 615–27.

Knill-Jones, R.P., Newman, B.J., and Spence, A.A. (1975). Anaesthetic practice and pregnancy. Controlled survey of male anaesthetists in the United Kingdom. *Lancet* **ii**, 807–9.

Koch, E.A., Smith, P.A., and King. R.C. (1967) The division and differentiation of *Drosophila* cystocytes. *J. Morph.* **121**, 55–70.

Koch, P., Pijnacker, L.P. and Kreke, J. (1972). DNA reduplication during meiotic prophase in the oocytes of *Carausius morosus* Br. *Chromosoma* **36**, 313–21.

Kostoff, D. (1938). Irregular mitosis and meiosis induced by acenaphtene. *Nature, Lond.* **141**, 1144.

—— (1948). Atypical growth, abnormal mitosis, polyploidy and chromosome fragmentation induced by hexachlorcyclohexane. *Nature, Lond.* **162**, 845–6.

Koulischer, L. and Gillerot, Y. (1980). Down's syndrome in Wallonia (South Belgium), 1971–1978. Cytogenetics and incidence. *Hum. Genet.* **54**, 243–50.

Kuleshov, N.P. (1976). Chromosome anomalies of infants dying during the perinatal period and premature newborn. *Humangenetik* **31**, 151–60.

Kusyk, C.J. and Hsu, T.C. (1976). Mitotic anomalies induced by three inhalation halogenated anesthetics. *Envir. Res.* **12**, 366–70.

Lajborek-Czyz, I. (1976). A 45,X woman with a 47,XY,G+ son. *Clin. Genet.* **9**, 113–16.

Lamy, R. (1949). Production of 2-X sperm in males. *Drosophila Inf. Serv.* **23**, 91.

Lander, E., Forssman, H., and Akesson, H.O. (1964). Season of birth and mental deficiency. *Acta genet.* **14**, 265–77.

Langenbeck, U., Hansmann, I., Hinney, B., and Hönig, V. (1976). On the origin of the supernumerary chromosome in autosomal trisomies—with special reference to Down's syndrome. *Hum. Genet.* **33**, 89–102.

Lauritsen, J.G. and Friedrich, J. (1976). Origin of the extra chromosome in trisomy 16. *Clin. Genet.* **10**, 156–60.

—— Jonasson, J., Therkelsen, A.J., Lass, F., Lindsten, J., and Petersen, G.B. (1972). Studies on spontaneous abortions. Fluorescence analysis of abnormal karyotypes. *Hereditas* **71**, 160–3.

Lawrence, C.W. (1961*a*). The effect of irradiation of different stages in microsporogenesis on chiasma frequency. *Heredity* **16**, 83–9.

—— (1961*b*). The effect of radiation on chiasma formation in *Tradescantia. Radiat. Bot.* **1**, 92–6.

—— (1965). The influence of non-lethal doses of radiation on recombination in *Chlamydomonas reinhardii. Nature, Lond.* **206**, 789–91.

—— (1968). Radiation effects on genetic recombination in *Chlamydomonas reinhardii.* In *Effects of radiation on meiotic systems,* pp.135–44. IAEA, Vienna.

Lea, D.E. (1955). *Action of radiation on living cells.* Cambridge University Press.

Leigh, B. (1979). Induced nondisjunction in *Drosophila* oocytes. *Mut. Res.* **61**, 65–8.

Lejeune, J., Gautier, M., and Turpin, R. (1959). Etude des chromosomes somatiques de neuf enfants mongoliens. *C. R. hebd. Séanc. Acad. Sci., Paris* **248**, 1721–2.

Lenz, W., Nowakowski, H., Prader, A., and Schirren, C. (1959). Die ätiologie des Klinefelter-syndroms. Ein beitrag zur chromosomen pathologie beim menschen. *Schweiz. med. Wochschr.* **89**, 727–31.

Levan, A. (1939*a*). Cytological phenomena connected with root swelling caused by growth substances. *Hereditas* **25**, 87–96.

—— (1939*b*). The effect of colchicine on meiosis in *Allium. Hereditas* **25**, 9–26.

—— and Ostergren, G. (1943). The mechanism of C-mitotic action observations on the naphthalene series. *Hereditas* **29**, 381–443.

—— and Tjio, J.H. (1948). Induction of chromosome fragments by phenols. *Hereditas* **34**, 453–84.

Levis, A.G. and Marin, G. (1963). Induction of multipolar spindles by X-radiation in mammalian cells *in vitro. Expl Cell Res.* **31**, 448–51.

Lewis, E.B. (1951). Additions and corrections to the cytology of rearrangements. *Drosophila Inf. Serv.* **25**, 108–9.

Lhoas, C. (1961). Mitotic haploidization by treatment of *Aspergillus niger* diploids with para-fluorophenylalanine. *Nature, Lond.* **190**, 744.

Licznerski, G. and Lindsten, J. (1972). Trisomy 21 in man due to maternal nondisjunction during the first meiotic division. *Hereditas* **70**, 153–4.

Lilienfeld, A.M. (1969) *Epidemiology of mongolism.* Johns Hopkins Press, Baltimore.

Lin, C.C. Gedeon, M.M., Griffith. P., Smink, W.K., Newton, D.R., Wilkie, L., and Sewell, L.M. (1976). Chromosome analysis on 930 consecutiv newborn children using quinacrine fluorescent banding technique. *Hum. Genet.* **31**, 315–28.

Lindsley, D.L. and Sandler, L. (1958). The meiotic behaviour of grossly deleted X-chromosomes in *Drosophila melanogaster. Genetics* **43**, 547–63.

Lindsten, J. (1963). *The nature and origin of X chromosome aberrations in Turner's syndrome. A cytogenetical and clinical study of 57 patients*. Almquist and Wiksell, Uppsala.

Liras, P., McKusker, J., Mascioli, S., and Haber, J.E. (1978). Characterization of a mutation in yeast causing non-random chromosome loss during mitosis. *Genetics* **88**, 651–71.

Lisker, R., Zenzes, M.T., and Fonesca, M.T. (1968). YY syndrome in a Mexican. *Lancet* **ii**, 635.

Long, S.E. and Williams, S.E. (1980). Frequency of chromosomal abnormalities in early embryos of the domestic sheep (*Ovis aries*). *J. Reprod. Fert.* **58**, 197–201.

Lubs, H.A. and Ruddle, F.H. (1970). Chromosomal abnormalities in the human population: estimation of rates based on New Haven newborn study. *Science, NY* **169**, 495–7.

Luning, K.G. (1982*a*). Genetics of inbred *Drosophila melanogaster*. VI. Crossing over in secondary nondisjunction exceptionals. *Hereditas* **96**, 161–74.

—— (1982*b*). Genetics of inbred *Drosophila melanogaster*. VII. Evidence of normal chromosome pairing in the distal ends of X-chromosomes in secondary nondisjunction. *Hereditas* **96**, 287–90.

Luthardt, F.W. (1977). Cytogenetic analyses of human oocytes. (Abstr.) *Am. J. hum. Genet.* **29**, 71A.

—— Palmer, C.G., and Yu, P.-L. (1973). Chiasma and univalent frequencies in ageing female mice. *Cytogenet. Cell Genet.* **12**, 68–79.

Luykx, P. (1970). Cellular mechanism of chromosomal distribution. *Int. Rev. Cytol.* Supple. **2**, 1–173.

Lycette, R.R., Whyte, S., and Chapman, C.J. (1970). Aneuploid effect of oestradiol on cultured human synovial cells. *NZ med. J.* **72**, 114–17.

Lyon, M.F. (1974). Sex chromosome activity in germ cells. In *Physiology and genetics of reproduction*, Part A (ed. E.M. Coutinho and F. Fuchs) pp.63–71. Plenum Press, New York.

—— and Hawker, S.G. (1973). Reproductive lifespan in irradiated and unirradiated chromosomally XO mice. *Genet. Res.* **21**, 185–94.

—— and Meredith, R. (1966). Autosomal translocations causing male sterility and viable aneuploidy in the mouse. *Cytogenetics* **5**, 335–54.

—— Ward, H.C., and Simpson, G.M. (1976). A genetic method for measuring nondisjunction in mice with Robertsonian translocations. *Genet. Res., Camb.* **26**, 283–95.

McClintock, B. (1933). The association of non-homologous parts of chromosomes in the midprophase of meioses in *Zea Mays*. *Z. Zellforsch. mikroskop. Anat. Abt. Histochem.* **19**, 191–237.

McDonald, A. (1972). Yearly and seasonal incidence of mongolism in Quebec. *Teratology* **6**, 1–4.

McFeely, R.A. (1967). Chromosome abnormalities in early embryos of the pig. *J. Reprod. Fert.* **13**, 579–81.

—— and Rajakoski, E. (1968). Chromosome studies on early embryos of the cow. *Proc. 6th Int. Cong. Anim. Reprod. and AI*, Vol. II. pp.905–7. Paris.

McGaughey, R.W. (1977). The culture of pig oocytes in minimal medium, and the influence of progesterone and estradiol-17β on meiotic maturation. *Endocrinology* **100**, 39–45.

—— and Chang, M.C. (1969). Inhibition of fertilization and production of heteroploidy in eggs of mice treated with colchicine. *J. exp. Zool.* **171**, 465–80.

McGill, M., Pathak, S., and Hsu, T.C. (1977). Effects of ethidium bromide on mitosis and chromosomes. A possible material basis for chromosome stickiness. *Chromosoma* **47**, 157–67.

Machin, G.A. and Crolla, J.A. (1974). Chromosome constitution of 500 infants dying during the perinatal period. *Humangenetik* **23**, 183–98.

McIntosh, J.R., Hepler, P.K., and van Wie, D.G. (1969). Model for mitosis. *Nature, Lond.* **224**, 659–63.

McQuarrie, H.G., Scott, C.D., Ellsworth, H.S., Harris, J.W., and Stone, R.A. (1970). Cytogenetic studies on women using oral contraceptives and their progeny. *Am. J. Obstet. Gynec.* **108**, 659–65.

Magenis, R.E., Hecht, F., and Milham, S. (1968). Trisomy 13 (D) syndrome: studies on parental age, sex ratio and survival. *J. Pediat.* **73**, 222–8.

—— Overton, K.M., Chamberlin, J., Brady, T., and Lovrien, E. (1977). Parental origin of the extra chromosome in Down's syndrome. *Hum. Genet.* **37**, 7–16.

Maguire, M.P. (1974). Chemically induced abnormal chromosome behaviour at meiosis in maize. *Chromosoma* **48**, 213–23.

—— (1982). Evidence for a role of the synaptonemal complex in provision for natural chromosome disjunction at meiosis II in maize. *Chromosoma* **84**, 675–86.

Malawista, S.E., Sato, H., and Bensch, K.G. (1968). Vinblastine and griseofulvin reversibly disrupt the living mitotic spindles. *Science, NY* **160**, 770–2.

Mangenot, G. and Carpentier, S. (1944). Sur les effects mitoclastiques de la caffeine et la theophylline. *C.r. Soc. Biol.* **138**, 232–3.

Mann, J.D., Cahan, A., Gelb, A.G., Fisher, N., Hamper, J., Tippett, P., Sanger, R., and Race, R.R. (1962). A sex-linked blood group. *Lancet* **i**, 8–10.

Mantel, N. and Stark, C.R. (1967). Paternal age in Down's syndrome. *Am. J. ment. Defic.* **71**, 1025–7.

Margolis, R.L. and Wilson, L. (1981). Microtubule treadmills—possible molecular machinery. *Nature, Lond.* **293**, 705–11.

Margulis, L., Neviackas, J.A., and Banerjee, S. (1969). Cilia regeneration in *Stentor*: inhibition delay and abnormalities induced by griseofulvin. *J. Protozool.* **16**, 660–7.

Martin, R.H., Dill, F.J., and Miller, J.R. (1976). Nondisjunction in ageing female mice. *Cytogenet. Cell Genet.* **17**, 150–60.

—— Lin, C.C., Balkan, W., and Burns, K. (1982). Direct chromosomal analysis of human spermatozoa: preliminary results from 18 normal men. *Am. J. hum. Genet.* **34**, 459–68.

Mason, J.M. (1976). Orientation disrupter (ord): a recombination-defective and disjunction-defective meiotic mutant in *Drosophila melanogaster*. *Genetics* **84**, 545–72.

Mattei, J.F., Mattei, M.G., Ayme, S., and Giraud, F. (1979). Origin of the extra chromosome in trisomy 21. *Hum. Genet.* **46**, 107–10.

Maudlin, I. and Fraser, L.R. (1977). The effect of PMSG dose on the incidence of chromosomal anomalies in mouse embryos fertilized *in vitro*. *J. Reprod. Fert.* **50**, 275–80.

—— —— (1978a). The effect of sperm and egg genotype on the incidence of chromosomal anomalies in mouse embryos fertilized *in vitro*. *J. Reprod. Fert.* **52**, 107–12.

—— —— (1978b). Maternal age and the incidence of aneuploidy in first-cleavage mouse embryos. *J. Reprod. Fert.* **54**, 423–6.

Mauer, I. (1963). Effect of hormones on mitosis and karyotype of human leukocytes cultured *in vitro*. *Genetics.* **48**, 899–900.

Mavor, J.W. (1921). On the elimination of the X-chromosome from the egg of *Drosophila melanogaster* by X-rays. *Science, NY* **54**, 277–9.

—— (1922). An effect of X-rays on inheritance. *Albany med. Ann.* 209–20.

—— (1923). Studies on the biological effects of X-rays. *Am. J. Roentgenol. radium Ther.* **X**, 968–74.

—— (1924). The production of nondisjunction by X-rays. *J. exp. Zool.* **39**, 381–432.

Max, C. (1977). Cytological investigation of embryos in low-dose X-irradiated young and old female inbred mice. *Hereditas* **85**, 199–206.

Mazia, D. (1958). SH compounds in mitosis I. The action of mercaptoethanol on the eggs of the sand dollar *Dendraster excentricus. Expl Cell Res.* **14**, 486–94.

Meisner, L.F., Chuprevich, T. W., Johnson, C.B., Inhorn, S,L., and Carter, J.J. (1973). Banding of human chromosomes with caesium chloride. *Lancet* **i**, 100–1.

Merriam, J.R. and Frost, J.N. (1964). Exchange and nondisjunction of the X-chromosomes in female *Drosophila melanogaster. Genetics* **49**, 109–22.

Michel, K.E. and Burnham, C.R. (1969). The behaviour of non-homologous univalents in double trisomics in maize. *Genetics* **63**, 851–64.

Mikamo, K. (1968). Mechanism of nondisjunction of meiotic chromosomes and of degeneration of maturation spindles in eggs affected by intrafollicular overripeness. *Experientia* **24**, 75–8.

—— (1979). Cytogenetic studies in meiotic and cleavage divisions on chromosomal nondisjunction. Int. Congress Series No. 512 Gynecology and Obstetrics. Proc. IX World Congress of Gynecology and Obstetrics, Tokyo (ed. S. Sakamoto, S. Tojo, and T. Nakayama). Excerpta Medica, Elsevier/North Holland, Amsterdam.

—— and Hamaguchi, H. (1975). Chromosomal disorder caused by pre-ovulatory overripeness of oocytes. In *Ageing gametes* (ed. R.J. Blandau) pp.72–97. Karger, Basel.

—— Aguercif, M., Hazeghi, P., and Martin-Du-Pain, R. (1968). Chromatin positive Klinefelter's syndrome: a quantitative analysis of spermatogonial deficiency at 3, 4 and 12 months of age. *Fert. Steril* **19**, 731–9.

Mikelsaar, A.-V., Schwarzacher, H.G., Schnedl, W., and Wagenbichler, P. (1977). Inheritance of Ag-stainability of nucleolus organizer regions. Investigations in 7 families with trisomy 21. *Hum. Genet.* **38**, 183–8.

Mikkelsen, M., Hallberg, A., and Poulsen, H. (1976). Maternal and paternal origin of extra chromosome in trisomy 21. *Hum. Genet.* **32**, 17–21.

—— Nielsen, G., and Rasmussen, E. (1978). Cost-effectiveness of antenatal screening for chromosome abnormalities. In *Towards the prevention of fetal malformation* (ed. J.B. Scrimgeour) pp.209–16. Edinburgh University Press.

—— Poulsen, H., Grinsted, J., and Lange, A. (1980). Nondisjunction in trisomy 21. Study of chromosomal heteromorphisms in 110 families. *Ann. hum. Genet.* **44**, 17–28.

Miklos, G.L.G. (1974). Sex chromosome pairing and male fertility. *Cytogenet. Cell Genet.* **13**, 558–77.

—— Yanders, A.F., and Peacock, W.J. (1972). Multiple meiotic drive systems in the *Drosophila melanogaster* male. *Genetics* **72**, 105–15.

Miller, J.F., Williamson, E., Glue, J., Gordon, Y.B., Grudzinskas, J.G., and Sykes, A. (1980). Fetal loss after implantation. Lancet **ii**, 554–6.

Miller, O.J., Breg, W.R., Schmickel, R.D., and Tretter, W. (1961). A family with an XXXXY male, a leukaemic male and two 21-trisomic mongoloid females. *Lancet* **ii**, 78–79.

Mir, L. and Wright, M. (1978). Action of anti microtubular drugs on *Physarum polycephalum. Microbiol. Lett.* **5**, 39–44.

Mirre, C., Hartung, M., and Stahl, A. (1980). Association of ribosomal genes in the fibrillar center of the nucleolus: a factor influencing translocation and nondisjunction in the human meiotic oocyte. *Proc. natn. Acad. Sci. USA* **77**, 6017–21.

Mitchell, M.B., Pittenger, T.H., and Mitchell, H.K. (1952). Pseudowild types in *Neurospora crassa. Proc. natn. Acad. Sci. USA* **38**, 569–80.

Mittler, S. (1976). Screening for induced chromosome loss and nondisjunction in *Drosophila melanogaster. Mut. Res.* **38**, 406–7.

—— Mittler, J.E., and Owens, S.L. (1967*a*). Loss of chromosomes and nondisjunction induced by caffeine in *Drosophila. Nature, Lond.* **214**, 424.

—— —— Tonetti, A.M., and Szymczak, M.E. (1967*b*). The effect of caffeine on chromosome loss and nondisjunction in *Drosophila. Mut. Res.* **4**, 708–9.

Mohr, O.L. (1919). Mikroskopische Untersuchungen zu Experimenten neber den Enfluss der Radiumstrahlen mol der Kaeltewirkung auf die chromatin reifung und das heterochromosom bei *Decticus verrucivorus. Arch. Mikroskop. Anat. Entewicklungsmech (Alb II)* **92**, 300–68.

Moor, R.M. (1978). Role of steroids in the maturation of ovine oocytes. *Ann. Biol. anim. Biochem. Biophys.* **18**, 477–82.

Moore, C.M. (1971). Non-homologous pairing in oogonia and ganglia of *Drosophila melanogaster. Genetica* **42**, 445–56.

Morpurgo, F., Bellincampi, D., Gualandi, G., Baldinelli, L., and Serlupi Crescenzi, O. (1979). Analysis of mitotic nondisjunction with *Aspergillus nidulans. Envir. Hlth Perspect.* **31**, 81–95.

Moustacchi, E., Hottinguer-de-Margerie, H., and Fabre, F. (1967). A novel character induced in yeast by P^{32} decay: the ability to manifest high frequencies of abnormal meiotic segregation. *Genetics* **57**, 909–18.

Mulcahy, M.T. (1978). Down syndrome and parental coital rate. *Lancet* ii, 895.

Müller, H.J. (1924). The nature of the genetic effects produced by radiation. In *Radiation biology* (ed. A. Hollaender) Part I, pp.351–473. McGraw Hill, New York.

—— (1925). The regionally differential effect of X-rays on crossing over in autosomes of *Drosophila. Genetics* **10**, 470–507.

—— (1927). Artificial transmutation of the gene. *Science, NY* **66**, 84–7.

Nagasawa, H. and Dewey, W.C. (1972). Effects of cold treatment on synchronous Chinese hamster cells treated in mitosis. *J. cell Physiol.* **80**, 89–106.

Nakashima, I. and Robinson, A. (1971). Fertility in a 45,X female. *Pediatrics* **47**, 770–3.

Nath, J. and Rebhun, L.I. (1976). Effects of caffeine and other methylxanthines on the development and metabolism of sea urchin eggs. Involvement of NADP+ and glutathione. *J. Cell Biol.* **68**, 440–50.

Nebel, B.R. and Hackett, E.M. (1961). Synaptinemal complexes (cores) in primary spermatocytes of mouse under elevated temperature. *Nature, Lond.* **190**, 467–8.

Nelson, J.F., Felicio, L.S., and Finch, C.E. (1981). Preovulatory rise of plasma estradiol is delayed in ageing mice with prolonged estrous cycles. *Biol. Reprod.* **24**, 784–94.

Nes, N. (1968). Betydningen av Kromosom aberrasjoner hos dyr. *Forskning og forsøk i Landbruket* **19**, 393–440.

Nicklas, R.B. (1967). Chromosome micromanipulation II. Induced reorientation and the experimental control of segregation in meiosis. *Chromosoma* **21**, 17–50.

—— and Staehly, C.A. (1967). Chromosome micromanipulation I. The mechanics of chromosome attachment to the spindle. *Chromosoma* **21**, 1–16.

Nicoletti, B, Trippa. G., and Demarco, A. (1967). Reduced fertility in SD males and its bearing on segregation distortion in *Drosophila melanogaster. Atti Accad. naz. Lincei Rc. Sed. solen.* **43**, 383–92.

Nielsen, J. and Friedrich, U. (1969). Seasonal variation in nondisjunction of sex chromosomes. *Humangenetik* **8**, 258–60.

—— and Sillesen, I. (1975). Incidence of chromosome aberrations among 11,148 newborn children. *Humangenetik* **30**, 1–12.

—— Friedrich, U., Hreidarsson, A., and Zeuthen, E. (1974). Frequency of 9qh+ and risk of chromosome aberrations in the progeny of individuals with 9qh+ *Hum. Genet* **21**, 211–16.

Niikawa, N., Merotto, E., and Kajii, T. (1977). Origin of acrocentic trisomies in spontaneous abortuses. *Hum. Genet.* **40**, 73–8.

Nokkalo, S. and Puro, J. (1976). Cytological evidence for a chromocenter in *Drosophila. Hereditas* **83**, 265–8.

Norby, D.E., Hegreberg, G.A., Thuline, H.C., and Findley, D. (1974). An XO cat. *Cytogenet. Cell Genet.* **13**, 448–53.

Nordensen, I. (1979). Population studies in northern Sweden. IX. Incidence of Down's syndrome by time, region and maternal age. *Hereditas* **91**, 257–62.

Novitski, E. (1964). An alternative to the distributive pairing hypothesis in *Drosophila. Genetics* **50**, 1449–51.

—— and Puro, J. (1978). A critique of theories of meiosis in the female of *Drosophila melanogaster. Hereditas* **89**, 51–67.

—— Peacock, W.J., and Engel, J. (1965). Cytological basis of 'sex ratio' in *Drosophila pseudobscura. Science, NY* **148**, 516–17.

Noyes, R.W. (1970). Physiology of ovarian ageing. In *Down's syndrome (mongolism)* (ed. V. Apgar). *Ann. NY Acad. Sci.* **171**, 517–25.

Nur, U. (1963). Meiotic pathenogenesis and heterochromatization in a soft scale, *Pulvinaria hydrangeae* (Coccoidea: Homoptera). *Chromosoma* **14**, 123–39.

Nybom, N. and Knutson, B. (1947). Investigations on C-mitosis in *Allium cepa*. I. The cytological effect of hexaclorcyclohexan. II. The cytological effect of vitamin K. *Hereditas* **33**, 220–34.

Oakberg, E.F. (1960). Irradiation damage to animals and its effect on their reproductive capacity. *J. dairy Sci.* **43**, Suppl. 54–67.

—— and DiMinno, R.L. (1960). X-ray sensitivity of primary spermatocytes of the mouse. *Int. J. rad. Biol.* **2**, 196–209.

Ohama, K. and Kajii, T. (1972). Monosomy 21 in spontaneous abortus. *Humangenetik* **16**, 267–70.

Ohno, S., Kaplan, W.D., and Kinosita, R. (1959). Do XY– and O-sperm occur in *Mus musculus? Expl Cell Res.* **18**, 382–4.

Oksala, T. (1958). Chromosome pairing, crossing over and segregation in meiosis in *Drosophila melanogaster. Cold Spring Harbor Symp. Quant. Biol.* **23**, 197–210.

Onfelt, A. and Ramel, C. (1979). Some aspects on the organization of microfilaments and microtubules in relation to nondisjunction. *Envir. Hlth Perspect.* **31**, 45–52.

Ord, M.G. and Stocken, L.A. (1966). Metabolic properties of histones from rat liver and thymus gland. *Biochem. J.* **98**, 888–97.

Ostergren, G. (1944). An efficient chemical for the induction of sticky chromosomes. *Hereditas* **30**, 213–16.

—— (1947). Heterochromatic B-chromosomes in *Anthoxanthum. Hereditas* **33**, 261–96.

—— (1951). Narcotized mitosis and the precipitation hypothesis of narcosis. In *Méchanisme de la narcose. Colloques int. Cent. natn. Rech. sci.* **26**, 77–88.

—— (1954). Polyploids and aneuploids of *Crepis capillaris* produced by treatment with nitrous oxide. *Genetica* **27**, 54–64.

—— (1957). Production of polyploids and aneuploids of *Phalaris* by means of nitrous oxide. *Hereditas* **43**, 512–16.

—— and Levan, A. (1943). The connection between c-mitotic activity and water solubility in some monocyclic compounds. *Hereditas* **29**, 496–8.

Ostertag, W. and Haake, J. (1966). The mutagenicity in *Drosophila melanogaster* of caffeine and of other compounds which produce chromosome breakage in human cells in culture. *Z. Vererbungslehre* **98**, 299–308.

Owellen, R.J., Owens, A.H., and Donigian, D.W. (1972). The binding of vincristine, vinblastine and colchicine to tubulin. *Biochem. Biophys. Res. Comm.* **47**, 685–91.

Paget, G.E. and Walpole. A.L. (1958). Some cytological effects of griseofulvin. *Nature, Lond.* **182**, 1320–1.

Paris Conference (1971). Standardization in human cytogenetics. *Birth Defects: Original Article Series* **VIII**, 7, 1972. The National Foundation, New York.

Parker, D.R. (1969). Heterologous interchange at meiosis in *Drosophila*. II. Some disjunctional consequences of interchanges. *Mut. Res.* **7**, 393–407.

—— and Busby, N. (1973). Observations concerning the effects on radiation on the segregation of chromosomes. *Mut. Res.* **18**, 33–46.

—— and Williamson, J.H. (1974). Some radiation effects on segregation in *Drosophila*. *Genetics* **78**, 163–71.

—— —— (1976). Aberration induction and segregation in oocytes. In *The genetics and biology of* Drosophila (ed. M. Ashburner and E. Novitski) Vol. 1c, pp.1251–68. Academic Press, New York.

—— —— and Gavin, J. (1974) The nature and time of occurrence of radiation induced nondisjunction of the acrocentric X and 4th chromosomes in *Drosophila melanogaster* females. *Mut. Res.* **24**, 135–48.

Parmentier, R. and Dustin, P. (1948). Early effects of hydroquinone on mitosis. *Nature, Lond.* **161**, 527–8.

—— —— (1953). On the mechanism of the mitotic abnormalities induced by hydroquinone in animal tissues. *Revue belge Path. Méd. exp.* **23**, 20–30.

Parry, J.M. (1977). The detection of chromosome nondisjunction in the yeast *Saccharomyces cerevisiae*. In *Progress in genetic toxicology* (ed. D. Scott, B.A. Bridges, and F.H. Sobels) pp.223–9. Elsevier/North Holland, Amsterdam.

—— Sharp, D., and Parry, E.M. (1979*a*). Detection of mitotic and meiotic aneuploidy in the yeast *Saccharomyces cerevisiae*. *Environ. Hlth Perspect.* **31**, 97–111.

—— —— Tippins, T.S., and Parry, E.M. (1979*b*). Radiation induced mitotic and meiotic aneuploidy in the yeast *Saccharomyces cerevisiae*. *Mut. Res.* **61**, 37–55.

Patau, K.A., Smith, D.W., Therman, E.M., Inhorn, S.L., and Wagner, H.P. (1960). Multiple congenital anomaly caused by an extra autosome. *Lancet* **i**, 790–3.

Pathak, S. and Hsu, T.C. (1977). Monitoring the effects of chemical mutagens with mammalian meiotic cells. *Mammal. Chrom. Newsl.* **18**, 42.

Patterson, J.T., Brewster, W., and Winchester, A.M. (1932). Effects produced by ageing and X-raying eggs of *Drosophila melanogaster*. *J. Hered.* **23**, 325–33.

Pawlowitski, I.H. and Pearson, P.L. (1972). Chromosomal aneuploidy in human spermatozoa. *Humangenetik* **16**, 119–22.

Peacock, W.J. (1965). Non-random segregation of chromosomes in *Drosophila* males. *Genetics* **51**, 573–83.

—— Miklos, G.L.G., and Goodchild, D.J. (1975). Sex chromosome meiotic drive systems in *Drosophila melanogaster* I. Abnormal spermatid development in males with a heterochromatin deficient X-chromosome. *Genetics* **79**, 613–34.

Pearson, P.L. (1972). The use of new staining techniques for human chromosome identification. *J. med. Genet.* **9**, 264–75.

—— and Bobrow, M. (1970). Fluorescent staining of the Y chromosome in meiotic stages of the human male. *J. Reprod. Fert.* **22**, 177–9.

Pena, A de la and Puertas, M.J. (1978). Colchicine-induced asynapsis and C-meiosis in pollen mother cells of cultivated anthers of rye. *Chromosoma* **68**, 261–7.

—— —— and Merino, F. (1981). Bimeiosis induced by caffeine. *Chromosoma* **83**, 241–8.

Penchaszadeh, V.B. and Coco, R. (1975). Trisomy 22. Two new cases and delineation of the phenotype. *J. med. Genet.* **12**, 193–9.

Penrose, L.S. (1933). The relative effects of paternal and maternal age in mongolism. *J. Genet.* **27**, 219–24.

—— (1934). The relative aetiological importance of birth order and maternal age in mongolism. *Proc. R. Soc. B* **115**, 431–43.

—— (1965). Mongolism as a problem in human biology. In *The early conceptus, normal and abnormal.* Papers and discussions presented at a Symposiun held at Queens College, Dundee, 1964, pp.94–7. University of St Andrews.

—— and Delhanty, J.D.A. (1961). Triploid cell cultures from a macerated foetus. *Lancet* **i**, 1261–2.

—— and Smith, G.F. (1966). *Down's anomaly.* Churchill, London.

Pera, F. and Schwarzacher, H.G. (1969). Die Verteilung der Chromosomen auf die Tochterzellkerne Mitosen euploiden Gewebekultusen, von *Microtus agrestis. Chromosoma* **26**, 337–54.

Peters, H., Byskov, A.G., and Grinsted, J. (1981). The development of the ovary during childhood in health and disease. In *Functional morphology of the human ovary* (ed. J.R.T. Coutts) pp. 26–34. MTP. Lancaster.

Petrova, L.G. (1976). Influence of ethylmethanesulphonate on nondisjunction and loss of X-chromosomes induced by X-rays in the germ cells of *Drosophila* females. *Soviet Genet.* **12**, 714–17.

Pharoah, P.O.D., Alberman, E., Doyle, P., and Chamberlain, G. (1977). Outcome of pregnancy among women in anaesthetic practice. *Lancet* **i**, 34–6.

Piko, L. and Bomsel-Helmreich, O. (1960). Triploid rat embryos and other chromosomal deviants after colchicine treatment and polyspermy. *Nature, Lond.* **186**, 737–9.

Pittenger, T.H. and Coyle, M.B. (1963). Somatic recombination in pseudowild-type cultures of *Neurospora crassa. Proc. natn. Acad. Sci. USA* **49**, 445–51.

Plough, H.H. (1917). The effect of temperature on crossing-over in *Drosophila. J. exp. Zool.* **24**, 147–209

Polani, P.E. (1961). Paternal and maternal nondisjunction in the light of colour vision studies. In *Human chromosomal abnormalities* (ed. W.M. Davidson and D. Robertson Smith) pp.80–3. Staples Press, London.

—— (1972). Centromere localization at meiosis and the position of chiasmata in the male and female mouse. *Chromosoma* **36**, 343–74.

—— and Jagiello, G.M. (1976). Chiasmata, meiotic univalents and age in relationship to aneuploid imbalance in mice. *Cytogenet. Cell Genet.* **16**, 505–29.

—— Briggs, J.H., Ford, C.E., Clarke, C.M., and Berg, J.M. (1960). A mongol girl with 46 chromosomes. *Lancet* **i**, 721–4.

Policansky, D. and Ellison, J. (1970). Sex ratio in *Drosophila pseudobscura*: spermiogenic failure. *Science, NY* **169**, 888–9

Purnell, D.J. (1973). Spontaneous univalence at male meiosis in the mouse. *Cytogenet. Cell Genet.* **12**, 327–35.

Puro, J. (1978). Recovery of radio-induced autosomal chromatid interchanges in oocytes of *Drosophila melanogaster*. *Hereditas* **88**, 203–11.
—— and Nokkala, S. (1977). Meiotic segregation of chromosomes in *Drosophila melanogaster* oocytes. A cytological approach. *Chromosoma* **63**, 273–86.
Quinlan, R.A., Roobol, A., Pogson, C.I., and Gull, K. (1981). A correlation between *in vivo* and *in vitro* effects of the microtubule inhibitors colchicine, parbendazole and nocodazole on myxamoebae of *Physarum polycephalum*. *J. gen. Microbiol.* **122**, 1–6.
Race, R.R. and Sanger, R. (1969). Xg and sex chromosome abnormalities. *Br. med. Bull.* **25**, 99–103.
Ramel, C. (1962). Interchromosomal effects of inversions in *Drosophila melanogaster*. II. Non-homologous pairing and segregation. *Hereditas* **48**, 59–82.
—— (1969). Genetic effects of organic mercury compounds. I. Cytological investigations on *Allium* roots. *Hereditas* **61**, 208–30.
—— and Magnusson, J. (1969). Genetic effects of organic mercury compounds. II. Chromosomal segregations in *Drosophila melanogaster*. *Hereditas* **61**, 231–54.
—— —— (1979). Chemical induction of nondisjunction in *Drosophila*. *Environ. Hlth Perspect.* **31**, 59–66.
Rao, P.N. and Engelberg, J. (1966). Mitotic nondisjunction of sister chromatids and anomalous mitosis induced by low temperatures in HeLa cells. *Expl Cell Res.* **43**, 332–42.
—— —— (1967). Structural specificity of estrogens in the induction of mitotic chromatid nondisjunction in HeLa Cells. *Expl Cell Res.* **48**, 71–81.
Rapaport, I. (1963). Oligophrénia mongolienne et caries dentaires. *Revue Stomatol., Paris* **64**, 207–10.
Rapoport, J.A. (1938). The effect of X-rays on nondisjunction of the fourth and X-chromosomes in *Drosophila melanogaster*. *Biologicheskii Zhurnal* **7**, 661–78.
Ratcliffe, S.G., Axworthy, D., and Ginsborg, A. (1979). The Edinburgh study of growth and development in children with sex chromosome abnormalities. In *Sex chromosome aneuploidy: prospective studies on children* (ed. A. Robinson, H.A. Lubs, and D. Bergsma). *Birth Defects: Original Article Series*, Vol. XV, No.1, pp. 243–60. The National Foundation.
—— Bancroft, J., Axworthy, D., and McLaren, W. (1982). Klinefelter's syndrome in adolescence. *Archs Dis. Child.* **57**, 6–12.
Raybin, D. and Flavin, M. (1975). An enzyme tyrosylating α-tubulin and its role in microtubule assembly. *Biochem Biophys. Res. Comm.* **65**, 1088–95.
—— (1977). Enzyme which specifically adds tyrosine to the α chain of tubulin. *Biochemistry* **16**, 2189–93.
Reichert, W., Hansmann, I., and Röhrborn, G. (1975). Chromosome anomalies in mouse oocytes after irradiation. *Humangenetik* **28**, 25–38.
Riccardi, V.M. (1977). Trisomy 8: an international study of 70 patients. The National Foundation – March of Dimes. *Birth Defects: Original Article Series*, Vol. XIII, No.3C, pp.171–84.
Richards, O.W. (1938). Colchicine stimulation of yeast growth fails to reveal mitosis. *J. Bact.* **36**, 187–95.
Rieck, G.W. (1970). The XXY syndrome in cattle (bovine hypogonadism). *Giessener Beitr. Erbpath Zuchthyg.* Suppl. **1**, 138–45.
—— Höhn, H., and Herzog, A. (1970). X-trisomie beim Rind mit Anzeichem familiäer Disposition fur Meiosestorüngen. *Cytogenetics* **9**, 401–9.
Roberts, A.M. and Goodall, H. (1976). Y-chromosome visibility in quinacrine-stained human spermatozoa. *Nature, Lond.* **262**, 493–4.

Robinson, A. (1974). Neonatal deaths and sex chromosome anomalies. *Lancet* **i**, 1223.

—— (1979). In *Sex chromosome aneuploidy: prospective studies on children. Birth Defects: Original Article Series*, Vol XV, No.1., pp.1–2. The National Foundation.

—— Lubs, H.A., Nielsen, J., and Sørensen, K. (1979). Summary of clinical findings: profiles of children with 47,XXY, 47,XXX and 47,XYY karyotypes. In *Sex chromosome aneuploidy: prospective studies on children. Birth Defects: Original Article Series*, Vol. XV, No.1, pp.261–6. The National Foundation.

Robinson, J.A. (1973). Origin of extra chromosome in trisomy 21. *Lancet* **i**, 131–3.

—— (1977). Meiosis I nondisjunction as the main cause of trisomy 21. *Hum. Genet.* **39**, 27–30.

—— and Newton, M. (1977). A fluorescence polymorphism associated with Down's syndrome. *J. med. Genet.* **14**, 40–5.

Rodman, T.C. (1971). Chromatid disjunction in unfertilized ageing oocytes. *Nature, Lond.* **233**, 191–3.

—— Flehinger, B.J. and Rohlf, F.J. (1980). Metaphase chromosome associations: colcemid distorts the pattern. *Cytogenet. Cell Genet.* **27**, 98–110.

Rodriguez, J.A. and Borisy, G.G. (1978). Modification of the C-terminus of brain tubulin during development. *Biochem. Biophys. Res. Comm.* **83**, 579–86.

—— —— (1979). Experimental phenylketonuria: replacement of carboxyl terminal tyrosine by phenylalanine in infant rat tubulin. *Science, NY* **206**, 463–5.

Röhrborn, G. and Hansmann, I. (1971). Induced chromosome aberrations in unfertilized oocytes of mice. *Hum. Genet.* **13**, 184–98.

—— —— (1974). Oral contraceptives and chromosome segregation in oocytes of mice. *Mut. Res.* **26**, 535–44.

Rondanelli, G.G., Trenta, A., Magliulo, E., Vannini, V., Gerna, G., and Carosi, G. (1967). Morphogénèse des cellules à plusieurs noyaux par l'action des rayons X. Etude en contraste de phase sur les cellules érythropoïétiques à l'etant vivant. *Expl Cell Res.* **47**, 222–8.

Roobol, A., Gull, K., and Pogson, C.I. (1976). Inhibition by griseofulvin of microtubule assembly *in vitro*. FEBS Lett. **67**, 248–51.

Roth, D.B. (1963). The frequency of spontaneous abortion. *Int. J. Fert.* **8**, 431–4.

Roth, L.E. (1967). Electron microscopy of mitosis in ameobae. III. Cold and urea treatments; a basis for tests of direct effects of mitotic inhibitors on microtubule formation. *J. cell Biol.* **34**, 47–59.

Rudak, E., Jacobs, P.A. and Yanagimachi, R. (1978). Direct analysis of the chromosome constitution of human spermatozoa. *Nature, Lond.* **274**, 911–13.

Rundle, A., Coppen, A., and Cowie, V. (1961). Steroid excretion in mothers of mongols. *Lancet* **ii**, 846–8.

Russell, L.B. (1961). Genetics of mammalian sex chromosomes. *Science, NY* **133**, 1795–803.

—— (1968). The use of sex-chromosome anomalies for measuring radiation effects in different germ-cell stages of the mouse. In *Effects of radiation on meiotic systems*, pp.27–41. IAEA, Vienna.

—— (1976). Numerical sex-chromosome anomalies in mammals: their spontaneous occurrence and use in mutagenesis studies. In *Chemical mutagens*, Vol. 4 (ed. A. Hollaender) pp. 55–91. Plenum Press, New York.

—— (1979). Meiotic nondisjunction in the mouse: methodology for genetic testing and comparison with other methods. *Envir. Hlth Perspect.* **31**, 113–28.

—— and Chu, E.H.Y. (1961). An XXY male in the mouse. *Proc. natn. Acad. Sci. USA* **47**, 571–5.

—— and Montgomery, C.S. (1974). The incidence of sex-chromosome anomalies following irradiation of mouse spermatogonia with single or fractionated doses of X-rays. *Mut. Res.* **25**, 367–76.

—— and Saylors, C.L. (1962). Induction of paternal sex-chromosome losses by irradiation of mouse spermatozoa. *Genetics* **47**, 7–10.

—— —— (1963). The relative sensitivity of various germ-cell stages of the mouse to radiation-induced nondisjunction, chromosome losses and deficiencies. In *Repair from genetic radiation damage* (ed. F.H. Sobels) pp.313–52. Pergamon Press, Oxford.

Russell, P. and Altschuler, G. (1975). The ovarian dysgenesis of trisomy 18. *Pathology* **7**, 149–55.

Ryan, T.J., Boddington, M.M., and Spriggs, A.J. (1965). Chromosomal abnormalities produced by folic acid antagonists. *Br. J. Derm.* **77**, 541–55.

Sadovnick, A.D. and Baird, P.A. (1981). A cost-benefit analysis of prenatal detection of Down syndrome and neural tube defects in older mothers. *Am. J. med. Genet.* **10**, 367–78.

Safir, S.R. (1920). Genetic and cytological examination of the phenomenon of primary nondisjunction in *Drosophila melanogaster*. *Genetics* **5**, 459–87.

Sandler, L. and Braver, G. (1954). A meiotic loss of unpaired chromosomes in *Drosophila melanogaster*. *Genetics* **39**, 365–77.

—— Lindsley, D.D., Nicoletti, B., and Trippa, G. (1968). Mutants affecting meiosis in natural populations of *Drosophila melanogaster*. *Genetics* **60**, 525–58.

Sanger, R., Tippett, P., Gavin, J. (1971). Xg groups and sex abnormalities in people of northern European ancestry. *J. med. Genet.* **8**, 417–26.

—— —— —— Teesdale, P., and Daniels, G.L. (1977). Xg groups and sex chromosome abnormalities in people of northern European ancestry: an addendum. *J. med. Genet.* **14**, 210–13.

Sansome, E.R. and Bannon, L. (1946). Colchicine ineffective in inducing polyploidy in *Penicillium notatum*. *Lancet* **ii**, 828–9.

Sasaki, M. (1965). Meiosis in a male with Down's syndrome. *Chromosoma* **16**, 652–7.

Sass, J.E. (1937). Histological and cytological studies of ethyl mercury phosphate. *Phytopathology* **27**, 95–9.

Sävhagen, R. (1960). Relation between X-ray sensitivity and cell stages in males of *Drosophila melanogaster*. *Nature, Lond.* **188**, 429–30.

Savontaus, M.-L. (1975). Relationship between effects of X-rays on nondisjunction and crossing over in *Drosophila melanogaster*. *Hereditas* **80**, 195–204.

—— (1977). X-ray induced autosomal nondisjunction in relation to crossing over in *Drosophila melanogaster*. *Hereditas* **86**, 115–20.

Schiavi, R.C., Owen, D., Fogel, M., White, D., and Szechter, R. (1978). Pituatory–gonadal function in XYY and XXY men identified in a population survey. *Clin. Endocr.* **9**, 233–9.

Schiff, P.B., Fant, J., and Horwitz, S.B. (1979). Promotion of microtubule assembly *in vitro* by taxol. *Nature, Lond.* **277**, 665–7.

Schleiermacher, E. (1968). Chemical mutagenesis in mammals and man. II. Cytological effects of radiomimetic drugs in meiosis in the male mouse. Proc. Int. Symp. on genetic effects of radiation and radiomimetic chemicals. *Jap. J. Genet.* **44**, Suppl. 2, 80–1.

Schnedl, W. (1971). Unterschiedliche Fluorescenz der beiden homologen chromosomen Nr. 3 beim menschen. *Humangenetik* **12**, 59–63.

Schneider, E.L. and Kram, D. (1981). Animal models for studying parental age effects. In *Trisomy 21 (Down syndrome)* (ed. F.F. de la Cruz and P.S. Gerald) pp.275–80. University Park Press, Baltimore.

Schull, W.J. and Neel, J.V. (1962). Maternal radiation and mongolism. *Lancet* i, 537.

Seabright, M., Gregson, N., and Mould, S. (1976). Trisomy 9 associated with an enlarged 9qh segment in a liveborn. *Hum. Genet.* **34**, 323–5.

Searle, A.G. (1975). Radiation-induced chromosome damage and the assessment of genetic risk. In *Modern trends in human genetics*, Vol. 2 (ed. A.E.H. Emery) pp.83–110. Butterworths, London.

Seiler, J.P. (1975). Toxicology and genetic effects of benzímidazole compounds. *Mut. Res.* **32**, 151–68.

—— (1976). Mutagenicity of benzimidazole and benzimidazole derivatives. VI. Cytogenetic effects of benzimidazole derivatives in the bone marrow of the mouse and Chinese hamster. *Mut. Res.* **40**, 339–48.

Sergovich, F., Valentine, G.H., Chen, A.T.L., Kinch, R.A.H., and Smout, M.S. (1969). Chromosome aberrations in 2159 consecutive newborn babies. *New Engl. J. Med.* **280**, 851–5.

Shanfield, B. and Käfer, E. (1971). Chemical induction of mitotic recombination in *Aspergillus nidulans. Genetics* **67**, 209–19.

Sharma, T. and Raman, R. (1971). An XO female in the Indian mole rat. *J. Hered.* **62**, 384–7.

Sharman, G.B., Robinson, E.S., Walton, S.M., and Berger, P.J. (1970). Sex chromosomes and reproductive anatomy of some intersexual marsupials. *J. Reprod. Fert.* **21**, 57–68.

Shepard, J., Boothroyd, E.R., and Stern, H. (1974). The effect of colchicine on synapsis and chiasma formation in microsporocytes of *Lilium. Chromosoma* **44**, 423–37.

Sherman, B.M., West, J.H., and Korenman, S.G. (1976). The menopausal transition: analysis of LH, FSH, estradiol and progesterone concentrations during menstrual cycles of older women. *J. Clin. Endocr. Metab.* **42**, 629–36.

Sigler, A.T., Lilienfeld, A.M., Cohen, B.H., and Westlake, J.E. (1965). Parental age in Down's syndrome (mongolism). *J. Pediat.* **67**, 631–42.

Simpson, J.L. (1976). *Disorders of sexual differentiation.* Academic Press, New York.

Singh, R.P. and Carr, D.H. (1966). The anatomy and histology of XO human embryos and fetuses. *Anat. Rec.* **155**, 369–83.

Sisken, J.E. and Iwasaki, T. (1969). The effects of some amino acid analogs on mitosis and the cell cycle. *Expl cell Res.* **55**, 161–7.

—— and Wilkes, E. (1967). The time of synthesis and the conservation of mitosis-related proteins in cultured human amnion cells. *J. cell Biol.* **34**, 97–110.

Skakkebaek, N.E., Hultén, M., and Philip, J. (1973a). Quantification of human seminiferous epithelium. IV. Histological studies in 17 men with numerical and structural autosomal aberrations. *Acta path. microbiol. scand.* **81**, 112–24.

—— Philip, J., and Hammen, R. (1969). Meiotic chromosomes in Klinefelter's syndrome. *Nature, Lond.* **221**, 1075–6.

—— Hultén, M., Jacobsen, P., and Mikkelsen, M. (1973b). Quantification of human seminiferous epithelium. II. Histological studies in eight 47,XYY men. *J. Reprod. Fert.* **32**, 391–401.

Smith, G.F. and Berg, J.M. (1976). *Down's anomaly*, 2nd edn. Churchill Livingstone, Edinburgh.

Smithers, D.W., Wallace, D.M., and Austin, D.E. (1973). Fertility after unilateral orchidectomy and radiotherapy for patients with malignant tumours of the testis. *Br. med. J.* **ii**,77–9.

Sobels, F.H. (1979). Studies of nondisjunction of the major autosomes in *Drosophila melanogaster*. II Effects of dose-fractionation, low radiation doses, EMS and ageing. *Mut. Res.* **59**, 179–88.

Sora, S., Lucchini, G., and Magni, G.E. (1982). Meiotic diploid progeny and meiotic nondisjunction in *Saccharomyces cerevisiae*. *Genetics* **101**, 17–33.

Speed, R.M. (1977). The effects of ageing on the meiotic chromosomes of male and female mice. *Chromosoma* **64**, 241–54.

—— and Chandley, A.C. (1981). The response of germ cells of the mouse to the induction of nondisjunction by X-rays. *Mut. Res.* **84**, 409–18.

—— —— (in press). Meiosis in the foetal mouse ovary. II. Oocyte development and age-related aneuploidy. Does a production line exist?

Spieler, R.A. (1963). Genic control of chromosome loss and nondisjunction in *Drosophila melanogaster*. *Genetics* **48**, 73–90.

Stark, C.R. and Mantel, N. (1967). Lack of seasonal- or temporal-spatial clustering of Down's syndrome births in Michigan. *Am. J. Epidemiol.* **86**, 199–213.

Stavrovskaya, A.A. and Kopnin, B.P. (1975). Colcemid-induced polyploidy and aneuploidy in normal and tumour cells *in vitro*. *Int. J. Cancer* **16**, 730–7.

Stearns, P.E., Droulard, K.E., and Sahhar, F.H. (1960). Studies bearing on fertility of male and female mongoloids. *Am. J. ment. Defic.* **65**, 37–41.

Stebbing, H. and Hyams, J.S. (1979). *Cell motility—integrated themes in biology*. Longman, London.

Stein, Z., Susser, M., and Guterman, A.V. (1973). Screening programme for prevention of Down's syndrome. *Lancet* **i**, 305–9.

Stene, J., Stene, E., Stengel-Rutkowski, S., and Murken, J-D. (1981). Paternal age and Down's syndrome. Data from prenatal diagnosis (DFG). *Hum. Genet.* **59**, 119–24.

—— Fischer, G., Stene, E., Mikkelsen, M., and Petersen, E. (1977). Paternal age effect in Down's syndrome. *Ann. hum. Genet.* **40**, 299–306.

Stern, C. (1926). An effect of temperature and age on crossing-over in the first chromosome of *Drosophila melanogaster*. *Proc. natn. Acad. Sci. USA* **12**, 530–2.

—— (1959). Colour blindness in Klinefelter's syndrome. *Nature, Lond.* **183**, 1452–3.

—— (1960). Addendum to Stewart—Mechanisms of meiotic nondisjunction in man. *Nature, Lond.* **187**, 805.

Stewart, J.S.S. (1960). Mechanisms of meiotic nondisjunction in man. *Nature, Lond.* **187**, 804–5.

—— and Sanderson, A.R. (1960). Fertility and oligophrenia in an apparent triple-X female. *Lancet* **ii**, 21–3.

Strangio, V.A. (1961). Radiosensitive stages in the spermatogenesis of *Drosophila melanogaster*. *Nature, Lond.* **192**, 781–2.

Strausmanis, R., Henrikson, I.-B., Holmberg, M., and Rönnbäck, C. (1978). Lack of effect on the chromosomal nondisjunction in aged female mice after low dose X-irradiation. *Mut. Res.* **49**, 269–74.

Strømnaes, O. (1968). Genetic changes in *Saccharomyces cerevisiae* grown on media containing DL-*p*-fluorophenylalanine. *Hereditas* **59**, 197–220.

Sturtevant, A.H. and Beadle, G.W. (1936). The relations of inversions in the X chromosome of *Drosophila melanogaster* to crossing over and nondisjunction. *Genetics* **21**, 554–604.

—— —— (1939). *An introduction to genetics*. Sanders, New York.

—— and Dobzhansky, T. (1936). Geographical distribution and cytology of 'sex ratio' in *Drosophila pseudobscura* and related species. *Genetics* **21**, 473–90.

Styles, J.A. and Garner, R. (1974). Benzimidazole carbamate methyl ester-evaluation of its effects *in vivo* and *in vitro*. *Mut. Res.* **26**, 177–87.

Sugawara, S. and Mikamo, K. (1980). An experimental approach to the analysis of mechanisms of meiotic nondisjunction and anaphase lagging in primary oocytes. *Cytogenet. Cell Genet.* **20**, 251–64.

—— —— (1983). Absence of correlation between univalent formation and meiotic nondisjunction in the aged female Chinese hamster. *Cytogenet. Cell Genet.* **35**, 34–40.

Sumner, A.T. and Robinson, J.A. (1976). A difference in dry mass between the heads of X- and Y-bearing human spermatozoa. *J. Reprod. Fert.* **48**, 9–15.

—— —— and Evans, H.J. (1971). Distinguishing between X, Y and YY-bearing human spermatozoa by fluorescence and DNA content. *Nature, New Biol.* **229**, 231–3.

Sutherland, G.R., Carter, R.F., and Morris, L.L. (1976). Partial and complete trisomy 9: delineation of a trisomy 9 syndrome. *Hum. Genet.* **32**, 133–40.

—— —— Bauld, R., Smith, I.I., and Bain, A.D. (1978). Chromosome studies at the paediatric necropsy. *Ann. hum. Genet.* **42**, 173–81.

Swanson, C.P., Merz, T., and Young, W.J. (1981). *Cytogenetics: the chromosome in division, inheritance and evolution*, 2nd edn. Prentice-Hall, Eaglewood Cliffs.

Szemere, G. (1978). Male meiotic and post-meiotic studies: a new possible way of mutagenicity testing. *Biol. Zbl.* **97**, 173–80.

—— and Chandley, A.C. (1975). Trisomy and triploidy induced by X-irradiation of mouse spermatocytes. *Mut. Res.* **33**, 229–38.

Täckholm, G. (1922). Zytologische Studien über die Gattung Rosa. *Acta hort. Berg* **7**, 97–381.

Takagi, N. and Sasaki, M. (1976). Digynic triploidy after superovulation in mice. *Nature, Lond.* **264**, 278–81.

Tarkowski, A. (1966). An air-drying method for chromosome preparation from mouse eggs. *Cytogenetics* **5**, 394–400.

Tates, A.D. (1979). *Microtus oeconomus* (Rodentia), a useful mammal for studying the induction of sex chromosome nondisjunction and diploid gametes in male germ cells. *Envir. Hlth Perspect.* **31**, 151–9.

—— and Vogel, N. de (1981). Further studies on effects of X-irradiation on prespermatid stages of the northern vole, *Microtus oeconomus*. Low induction of sex chromosomal nondisjunction and very high induction of diploid spermatids. *Mut. Res.* **2**, 323–30.

—— Pearson, P.L., and Geraedts, J.P.M. (1975). Identification of X and Y spermatozoa in the northern vole, *Microtus oeconomus*. *J. Reprod. Fert.* **42**, 195–8.

—— —— Ploeg, M.V.D., and Vogel, N. de (1979). The induction of sex-chromosomal nondisjunction and diploid spermatids following X-irradiation of pre-spermatid stages in the northern vole, *Microtus oeconomus*. *Mut. Res.* **61**, 87–101.

Taylor, A.I. (1968). Autosomal trisomy syndromes: a detailed study of 27 cases of Edward's syndrome and 27 cases of Patau's syndrome. *J. med. genet.* **5**, 227–52.

Tease, C. (1981). Chromosome nondisjunction in female mice after X-irradiation. *Mouse News Lett.* **65**, 22.

—— (1982). Similar dose-related chromosome nondisjunction in young and old female mice after X-irradiation. *Mut. Res.* **95**, 287–96.

Teplitz, R.L., Gustafson, P.E. and Pellet, O.L. (1968). Chromosome distribution in interspecific *in vitro* hybrid cells. *Expl Cell Res.* **52**, 379–91.

Tettenborn, U., Gropp, A., Mürken, J.D., Tinnefeld, W., Führmann, W., and Schwinger, E. (1970). Meiosis and testicular histology in XYY males. *Lancet* **ii**, 267–8.

Therkelsen, A.J., Grunnet, N., Hjort, T., Myhre Jensen, O., Jonasson, J., Lauritsen, J.G., Lindsten, J., and Bruun Petersen, G. (1973). Studies on spontaneous abortions. In *Chromosomal errors in relation to reproductive failure* (ed. A. Boué and C. Thibault) pp. 81–94. INSERM, Paris.

Thompson, H., Melnyk, J., and Hecht, F. (1967). Reproduction and meiosis in XYY. *Lancet* **ii**, 831.

Threlkeld, S.F.H. and Stephens, V. (1966). Ascospore isolates of *Neurospora crassa* giving rise to cultures containing two or more genetically different nuclei. *Can. J. Genet. Cytol.* **8**, 414–21.

—— and Stoltz, J.M. (1970). A genetic analysis of nondisjunction and mitotic recombination in *Neurospora crassa*. *Genet. Res.* **16**, 29–35.

Timson, J. (1969). Trisomy after colchicine therapy. *Lancet* **i**, 370.

Tokunaga, C. (1970*a*). The effects of low temperature and ageing on nondisjunction in *Drosophila*. *Genetics* **65**, 75–94.

—— (1970*b*). Aspects of low-temperature induced meiotic nondisjunction in *Drosophila* females. *Genetics* **66**, 653–61.

—— (1971). The effects of temperature and ageing of *Drosophila* males on the frequencies of XXY and XO progeny. *Mut. Res.* **13**, 155–61.

Traut, H. (1964). The dose-dependence of X-chromosome loss and nondisjunction induced by X-rays in oocytes of *Drosophila melanogaster*. *Mut. Res.* **1**, 157–62.

—— (1970). Nondisjunction induced by X-rays in oocytes of *Drosophila melanogaster*. *Mut. Res.* **10**, 125–132.

—— (1971). The influence of the temporal distribution of the X-ray dose on the induction of X-chromosomal nondisjunction and X chromosome loss in oocytes of *Drosophila melanogaster*. *Mut. Res.* **12**, 321–7.

—— (1978). The induction of X-chromosomal aneuploidy by 5-fluorodeoxyuridine (FUdR) fed to *Drosophila melanogaster* females. *Can. J. Genet. Cytol.* **20**, 259–63.

—— (1980). X-chromosomal nondisjunction induced by ageing oocytes of *Drosophila melanogaster*. The special susceptibility of mature eggs. *Can. J. Genet. Cytol.* **22**, 433–7.

—— and Scheid, W. (1971). The production of monosomic and trisomic individuals in *Drosophila melanogaster* by X-irradiation of immature oocytes. *Mut. Res.* **13**, 429–32.

—— —— (1974). The induction of aneuploidy by colcemid fed to *Drosophila melanogaster* females *Mut. Res.* **23**, 179–88.

—— and Schröder, A.J. (1978). The increase in the frequency of X-chromosomal aneuploidy in *Drosophila melanogaster* as a consequence of suppressed oviposition. *Mut. Res.* **49**, 225–32.

Treloar, A.E., Boynton, R.E., Behn, B.G., and Brown, B.W. (1967). Variation of the human menstrual cycle through reproductive life. *Int. J. Fert.* **12**, 77–126.

Tumba, A. (1974). L'influence de l'age parental sur la production de l'anomalie XXXXY. *J. Génét. hum.* **22**, 73–97.

Turner, H.H. (1938). A syndrome of infantilism, congenital webbed neck and cubitus valgus. *Endocrinology* **23**, 566–74.

Uchida, I. (1962). The effects of maternal age and radiation on the rate of nondisjunction in *Drosophila melanogaster*. *Can. J. Genet. Cytol.* **4**, 402–8.

—— (1970). Epidemiology of mongolism: the Manitoba study. In *Down's syndrome (mongolism)* (ed. V. Apgar) *Ann. NY Acad. Sci.* **171**, 361–9.

—— (1979). Radiation-induced nondisjunction. *Envir. Hlth Perspect.* **31** 13–18.

—— and Freeman, C.P. (1977). Radiation-induced nondisjunction in oocytes of aged mice. *Nature, Lond.* **265**, 186–7.

—— and Lee, C.P.V. (1974). Radiation-induced nondisjunction in mouse oocytes. *Nature, Lond.* **250**, 601–2.

—— —— and Byrnes, E.M. (1975). Chromosome aberrations induced *in vitro* by low doses of radiation: nondisjunction in lymphocytes of young adults. *Am. J. hum. Genet.* **27**, 419–29.

—— Ray, M., McRae, K.N., and Besant, D.F. (1968). Familial occurrence of trisomy 22. *Am. J. hum. Genet.* **20**,107–18.

Uretz, R.B. and Zirkle, R.E. (1955). Disappearance of spindles in sand dollar blastomeres after ultra-violet irradiation of cytoplasm. *Biol. Bull.* **109**, 370.

—— Bloom, W., and Zirkle, R.E. (1954). Irradiation of parts of individual cells. II. Effects of an ultraviolet microbeam focussed on parts of chromosomes. *Science, NY* **120**, 197–9.

Vaughan, M. and Steinberg, D. (1960). Biosynthetic incorporation of fluorophenylalanine into crystalline proteins. *Biochim. biophys. Acta* **40**, 230–6.

Verschaeve, L., Kirsch-Volders, M., Hens, L., and Susanne, C. (1978). Chromosome distribution studies in phenyl mercury acetate-exposed subjects and age-related controls *Mut. Res.* **57**, 335–47.

—— —— Susanne, C., Groetenbriel, C., Hamstermans, R., Lecomte, A., and Roosels, D. (1976). Genetic damage induced by occupational low mercury exposure. *Envir. Res.* **12**, 306–16.

—— Tassignon, J.-P., Lefevre, M., De Stoop, P., and Susanne, C. (1979). Cytogenetic investigation on leukocytes of workers exposed to metallic mercury. *Envir. Mutagen* **1**, 259–68.

Vickers, A.D. (1969). Delayed fertilization and chromosomal anomalies in mouse embryos. *J. Reprod. Fert.* **20**, 69–76.

Wagenbichler, P., Killian, W., Rett, A., and Schnedl, W. (1976). Origin of the extra chromosome No. 21 in Down's syndrome. *Hum. Genet.* **32**, 13–16.

Wahrman, J. and Fried, K. (1970). The Jerusalem prospective newborn survey of mongolism. *Ann. NY Acad. Sci.* **171**, 341–60.

Wald, N., Howard Turner, J., and Borges, W. (1970). Down's syndrome and exposure to X-irradiation. *Ann. NY Acad. Sci.* **171**, 454–66.

Walker, F.A. (1969). Trisomy after colchicine therapy. *Lancet* **i**, 257–8.

Walker, H.C. (1977). Comparative sensitivities of meiotic prophase stages in male mice to chromosome damage by acute X- and chronic gamma-irradiation. *Mut. Res.* **44**, 427–32.

Walzer, S. and Gerald, P.S. (1977). *Population cytogenetics. Studies in humans* (ed. E.B. Hook and I.H. Porter) pp.45–61. Academic Press, London.

Wang, R.W., Rebhun, L.I., and Kupchan, S.M. (1977). Antimitotic and antitubulin activity of the tumor inhibitor steganacin. *Cancer Res.* **37**, 3071–9.

Warburton, D. and Fraser, F.C. (1964). Spontaneous abortion risks in man: data from reproductive histories collected in a Medical Genetics Unit. *Am. J. hum. Genet.* **16**, 1–27.

—— Kline, J., Stein, Z., and Susser, M. (1980a). Monosomy X: a chromosomal anomaly associated with young maternal age. *Lancet* **i**, 167–9.

—— Stein, Z., Kline, J., and Susser, M. (1980b). Chromosome abnormalities in spontaneous abortion: data from the New York City study. In *Human embryonic and fetal death* (ed. I.H. Porter and E.B. Hook) pp.261–88. Academic Press, New York.

—— Yu, C., Kline, J., and Stein, Z. (1978). Mosaic autosomal trisomy in cultures from spontaneous abortions. *Am. J. hum. Genet.* **30**, 609–17.

Watanabe, T. and Endo, A. (1982). Chromosome analysis of preimplantation embryos after cadmium treatment of oocytes at meiosis I. *Environ. Mut.* **4**, 563–7.

——Shimada, T., and Endo, A. (1977). Stage specificity of chromosome mutagenicity induced by cadmium in mouse oocytes. *Teratology* **16**, 127.

—— —— —— (1979). Mutagenic effects of cadmium on mammalian oocyte chromosomes. *Mut. Res* **67**, 349–56.

Weber, D.F. (1969). A test of distributive pairing in *Zea mays*. *Chromosoma* **27**, 354–70.

Weber, W.W. (1967). Survival and the sex ratio in trisomy 17–18. *Am. J. hum. Genet.* **19**, 369–77.

Weinstein, A. (1936). The theory of multiple-strand crossing over. *Genetics* **21**, 155–99.

Weisenberg, R.C. (1972). Microtubule formation *in vitro* in solutions containing low calcium concentrations. *Science, NY* **177**, 1104–5.

Weiss, G., Weick, R.F., Knobil, E., Wolman, S.R., and Gorstein, F. (1973). An XO anomaly and ovarian dysgenesis in a rhesus monkey. *Folia primatol.* **19**, 24–7.

Welker, D.L. and Williams, K.L. (1980). Mitotic arrest and chromosome doubling using thiabendazole, cambendazole nocodazole and benlate in the slime mould *Dictyostelium discoideum*. *J. gen. Microbiol.* **116**, 397–407.

Welshons, W.J. and Russell, L.B. (1959). The Y-chromosome as the bearer of male-determining factors in the mouse. *Proc. natn. Acad. Sci. USA* **45**, 560–6.

Wennström, J. (1971). Effects of ionizing radiation on the chromosomes in meiotic and mitotic cells. *Commentat. Biol.* **45**, 1–60.

Went, M. (1966). An indirect method of assay for mitotic centers in sand dollar (*Dendraster excentricus*) eggs. *J. cell Biol.* **30**, 555–62.

Westerman, M. (1967). The effect of X-irradiation of male meiosis in *Schistocerca gregaria* (Forskål). *Chromosoma* **22**, 401–16.

Westhead, E.W. and Boyer, P.D. (1961). The incorporation of *p*-fluorophenylalanine into some rabbit enzymes and other proteins. *Biochim. biophys. Acta* **54**, 145–56.

Westra, A. and Dewey, W.C. (1971). Variation in sensitivity to heat shock during the cell cycle of Chinese hamster cells *in vitro*. *Int. J. radiat. Biol.* **19**, 467–77.

White, M.J.D. (1935). The effects of X-rays on mitosis in the spermatogonial division of *Locusta migratoria*. *Proc. R. Soc. B* **119**, 61–84.

—— (1937). The effect of X-rays on the first meiotic division in three species of Orthoptera. *Proc. R. Soc. B* **124**, 183–96.

—— (1973). *Animal cytology and evolution*, 3rd edn. Cambridge University Press.

Williams, D.L., Hagen, A.A., and Runyan, J.W. (1971). Chromosome alterations produced in germ cells of dog by progesterone. *J. Lab. clin. Med.* **77**, 417–29.

—— Runyan, J.W., and Hagen, A.A. (1972). Progesterone-induced alterations of oogenesis in the Chinese hamster. *J. Lab. clin. Med.* **79**, 972–7.

Williams, K.L. (1980). Examination of the chromosomes of *Polyspondium pallidum* following metaphase arrest by Benzimidazole derivatives and colchicine. *J. gen. Microbiol.* **116**, 409–15.

Williamson, E.M. and Miller, J.F. (1980). A prospective study into early conceptual loss. *Clin. Genet.* **17**, 93.

Witkin, H.A., Mednick, S.A., Schulsinger, F., Bakkestrøm, E., Christiansen, K.O., Goodenough, D.R., Hirschhorn, K., Lundsteen, C., Owen, D.R., Philip, J., Rubin, D.B., and Stocking, M. (1976). Criminality in XYY and XXY men. *Science, NY* **193**, 547–55.

Witschi, E. and Laguens, R. (1963). Chromosomal aberrations in embryos from overripe eggs. *Devl Biol.* **7**, 605–16.

Yamamoto, M. and Ingalls, T.H. (1972). Delayed fertilization and chromosome anomalies in the hamster embryo. *Science, NY* **176**, 518–21.

—— Endo, A., and Watanabe, G. (1973*a*). Maternal age dependence of chromosome anomalies. *Nature, New Biol.* **241**, 141–2.

—— Shimada, T., Endo, A., and Watanabe, G.-I. (1973*b*). Effects of low-dose X-irradiation on the chromosomal nondisjunction in aged mice. *Nature, New Biol.* **244**, 206–8.

Yanagimachi, R., Yanagimachi, H., and Rogers, B.J. (1976). The use of zona-free animal ova as a test-system for the assessment of the fertilizing capacity of human spermatozoa. *J. Biol. Reprod.* **15**, 471–6.

Yong, H.S. (1971). Presumptive X monosomy in black rats from Malaya. *Nature, Lond.* **232**, 484–5.

Yunis, E., Ramírez, E., and Uribe, J.G. (1980). Full trisomy 7 and Potter syndrome. *Hum. Genet.* **54**, 13–18.

Zackai, E., Aronson, M., Kohn, G., Moorhead, P., and Mellman, W. (1973). Familial trisomy 22. *Am. J. hum. Genet.* **25**, 89A.

Zarfas, D.E. and Wolf, L.C. (1979). Maternal age patterns and the incidence of Down's syndrome. *Am. J. ment. Defic.* **83**, 353–9.

Zartman, D.L., Hinesley, L.L., and Gnatowski, M.W. (1981). A 53,X female sheep (*Ovis aries*). *Cytogenet. Cell Genet.* **30**, 54–8.

Zettle, T.E. and Murnik, M.R. (1973). Effects of caffeine on chromosome loss and nondisjunction in *Drosophila melanogaster. Genetics* **44**, 146–53.

Zimmering, S. and Wu, C.K. (1963). Radiation induced X-Y exchange and nondisjunction in spermatocytes of the immature testis of *Drosophila. Genetics* **48**, 1619–23.

Zsako, S. and Kaplan, A.R. (1969). Titres of antistreptolysin O in mothers of children affected with Down's syndrome. *Nature, Lond.* **223**, 1281–2.

Zutsch, U. and Kaul, B.L. (1975). Studies on the cytogenetic activity of some common fungicides in higher plants. *Cytobios* **12**, 61–7.

Author index

Subject Index